ENVIRONMENTAL RESEARCH ADVANCES

LEACHATES

BIOLOGICAL TREATMENT, TOXICOLOGICAL CHARACTERISTICS AND POTENTIAL IMPACTS ON THE ENVIRONMENT

Environmental Research Advances

Additional books in this series can be found on Nova's website under the Series tab.

Additional e-books in this series can be found on Nova's website under the eBooks tab.

ENVIRONMENTAL RESEARCH ADVANCES

LEACHATES

BIOLOGICAL TREATMENT, TOXICOLOGICAL CHARACTERISTICS AND POTENTIAL IMPACTS ON THE ENVIRONMENT

MARIA R. MATTOX
EDITOR

nova
science publishers
New York

Library of Congress Cataloging-in-Publication Data

ISBN: 978-1-53612-508-5

Published by Nova Science Publishers, Inc. † New York

CONTENTS

PREFACE

In chapter one by Antoine P. Trzcinski, PhD, a simulated Organic Fraction of Municipal Solid Waste (OFMSW) was treated in an anaerobic two-stage membrane process in order to develop and optimize said process, effectively allowing for improvement upon the current performances obtained with single stage designs. In chapter two, Huu Tap Van and Van Tuyen Trinh present a study with the intent to improve current methods of removal of organic compounds from landfill leachate by Peozone. In chapter three by Hamidi Abdul Aziz and Hossein Farraji, main hazardous compounds of LL and their expected effects on the environment are introduced, as well as plant-based treatment methods. Next, chapter four by Adekunle A. Bakare, Chibuisi G. Alimba, Titilayo Hassan, Ebenezer O. Ajayi, and Akindayo A. Sowunmi explores the 96 h median lethal toxicity and cytogenotoxicity of a sawmill wood leachate from a sawmill dumpsite at Ibadan, Nigeria in Clarias gariepinus were investigated, providing further research on the DNA damaging potential of sawmill wood leachate (SWL) in aquatic vertebrates.

Chapter 1 - In this chapter, a simulated Organic Fraction of Municipal Solid Waste (OFMSW) was treated in an anaerobic two-stage membrane process. The OFMSW feedstock was fed to a ten litre hydrolytic reactor (HR) where solid and liquid fractions were separated by a coarse mesh, while the leachate was fed to a three litre submerged anaerobic membrane bioreactor (SAMBR) with *in situ* membrane cleaning by biogas sparging beneath a flat sheet microfiltration module. The aim was to develop and optimize this two-stage process where the use of a membrane in both reactors to uncouple the Solid and Liquid Retention Times (SRT and HRT) would allow us to improve the current performances obtained with single stage designs. The Denaturing Gradient Gel Electrophoresis (DGGE) technique was used to monitor the microbial population in the reactors and have a better understanding of the archaeal and bacterial distribution in a two-stage process. It was found that Chemical Oxygen Demand (COD) removal was greater than 90% at a Hydraulic Residence Time (HRT) of 1.6-2.3 days at a maximum Organic Loading Rate (OLR) of 20 g COD/L.day. Even though the influent COD of the leachate was constantly changing giving rise to a transitory Food to Microorganisms ratio over time, the permeate COD from the SAMBR was typically between 300 and 500 mg/L, which can therefore be defined as a stabilised leachate. Because of the fluctuating properties of the leachate produced in the HR, the process was deemed more representative because the SAMBR treated a leachate with varying organic strength which is what can happen on a full scale. The COD removal in SAMBR2 was 94.5% on average, and only 1.6% in the subsequent aerobic polishing bioreactor (AMBR-Aerobic Membrane Bioreactor), so that a total COD removal of 96.1% was achieved at 0.4 day HRT. On average, 26% of the recalcitrants from SAMBR2 could be degraded aerobically in the AMBR. However, as in SAMBR1, at HRTs lower than 2 days, particulate solids in the leachate built up at the bottom of the SAMBR, eventually leading to the diffuser blocking. At MLTSS beyond 20 g/L, the transmembrane pressure (TMP) culminated at 850 mbar and the flux dropped to 0.5 $L.m^{-2}.hr^{-1}$ LMH until the end of the experiment. The permeate of the SAMBR was low in COD and relatively constant which promoted the growth of autotrophic bacteria in the subsequent AMBR, so that 97.7% of the NH_4^+-N was removed at a maximum nitrogen loading rate of 0.18 kg NH_4^+-N/m^3.day. Calcium in the leachate was found to precipitate in the AMBR because of the higher pH. A sample taken from the membrane consisted most likely of pure hydroxylapatite $Ca_5(PO_4)_3(OH)$ which had a needle shape, whereas the background of the precipitate consisted more of nodules of calcium carbonate with traces of manganese, iron, magnesium, aluminium, sulphur and sodium. The very high molecular weight (MW) aromatic organics in the leachate fed to the SAMBRs were almost fully degraded in the bulk of the SAMBRs. Moreover, their permeate was absolutely free of them which indicates a full rejection of these compounds by the membrane. The medium MW compounds in the range of 395 - 646 kDa were more likely to be rejected by the membrane while the MW $\leq$ 395 kDa were observed in the permeate. Regarding the evolution of the medium and low MW compounds over the 200 days, it can be stated that overall there was no build-up in the absorbance in any effluent.

Chapter 2 - For improving the treatment of landfill leachate by combination O_3/H_2O_2 and Ceramic Raschig Rings (CRR) with different surface areas were added to enhance the mass transfer of ozone into liquid phase. To determine optimal conditions of reaction, pH and reaction time were also studied. The optimal pH ranges of 8-9, optimal reaction time of 80 min were identified in this research. CRR contribute to the significant improvement of the efficiency of landfill leachate treatment by O_3/H_2O_2 systems. With added CRR of $728m^2/m^3$ surface area, color, COD and TOC removal were increased in comparison with experiments without CRR was 8%, 14% and 9%, respectively. In this condition, the ozone utilization efficiency was also higher than that in experiments without CRR. The content of O_3 was also identified using 3.441 kg O_3/kg COD.

Chapter 3 - Landfill leachate (LL) is a high-strength wastewater and poses a potential environmental threat if not managed properly. High concentrations of hazardous heavy metals, phenols, and ammonia nitrogen are concerns in LL. Plant application in aquatic media as a sustainable and eco-friendly technique is a potential biological treatment method of LL. This chapter introduces the main hazardous compounds of LL and their expected effects on the environment. Plant-based treatment methods (phytoremediation and constructed wetland) for LL are also discussed.

Chapter 4 - With the rising global human population, trees are increasingly being cut to produce furniture, paper and building materials for human use. Common phenomena are the piling up of logs of wood and sawmills' shavings in the environment. The logs of wood and sawmills' shavings generate leachate that contain xenobiotics capable of contaminating aquatic and terrestrial ecosystems. They may also induce adverse effects on biota. Information is scarce on the DNA damaging potential of sawmill wood leachate (SWL) in aquatic vertebrates. Herein, the 96 h median lethal toxicity and cytogenotoxicity of a SWL from a sawmill dumpsite at Ibadan, Nigeria in *Clarias gariepinus* were investigated. Six concentrations (10 - 100%) of the leachate were examined for 96 h acute toxicity. Fish were exposed to sub-lethal concentrations (1.32 - 21.05%) of the 96 h LC_{50} for 7, 14 and 28 days to assess micronucleus (MN) and nuclear abnormalities in peripheral erythrocytes of five fish per concentration. The 96 h LC_{50} of 42.09% was 1.54 times more toxic than the 24 h LC_{50} (64.98%), indicating that the toxicity of the sawmill leachate increased with exposure time. The leachate induced significant ($p<0.05$) time-dependent increase in the frequency of MN and other nuclear abnormalities such as bleb, notch, bud and binucleated nuclei compared with the negative control. The concentrations of some of the SWL constituents were above permissible limit; and their interactions in the test organism provoked the observed mortality and genomic disruptions. These suggest that wood leachates can cause somatic mutation in fish.

In: Leachates
Editor: Maria R. Mattox
ISBN: 978-1-53612-508-5

Chapter 1

BIOLOGICAL TREATMENT OF LEACHATE IN A SUBMERGED ANAEROBIC MEMBRANE BIOREACTOR

Antoine P. Trzcinski, PhD
University of Southern Queensland, Toowoomba, Australia

ABSTRACT

In this chapter, a simulated Organic Fraction of Municipal Solid Waste (OFMSW) was treated in an anaerobic two-stage membrane process. The OFMSW feedstock was fed to a ten litre hydrolytic reactor (HR) where solid and liquid fractions were separated by a coarse mesh, while the leachate was fed to a three litre submerged anaerobic membrane bioreactor (SAMBR) with *in situ* membrane cleaning by biogas sparging beneath a flat sheet microfiltration module. The aim was to develop and optimize this two-stage process where the use of a membrane in both reactors to uncouple the Solid and Liquid Retention Times (SRT and HRT) would allow us to improve the current performances obtained with single stage designs. The Denaturing Gradient Gel Electrophoresis (DGGE) technique was used to monitor the microbial population in the reactors and have a better understanding of the archaeal and bacterial distribution in a two-stage process.

It was found that Chemical Oxygen Demand (COD) removal was greater than 90% at a Hydraulic Residence Time (HRT) of 1.6-2.3 days at a maximum Organic Loading Rate (OLR) of 20 g COD/L.day. Even though the influent COD of the leachate was constantly changing giving rise to a transitory Food to Microorganisms ratio over time, the permeate COD from the SAMBR was typically between 300 and 500 mg/L, which can therefore be defined as a stabilised leachate. Because of the fluctuating properties of the leachate produced in the HR, the process was deemed more representative because the SAMBR treated a leachate with varying organic strength which is what can happen on a full scale.

The COD removal in SAMBR2 was 94.5% on average, and only 1.6% in the subsequent aerobic polishing bioreactor (AMBR-Aerobic Membrane Bioreactor), so that a total COD removal of 96.1% was achieved at 0.4 day HRT. On average, 26% of the recalcitrants from SAMBR2 could be degraded aerobically in the AMBR. However, as in SAMBR1, at HRTs lower than 2 days, particulate solids in the leachate built up at the bottom of the SAMBR, eventually leading to the diffuser blocking. At MLTSS beyond 20 g/L, the transmembrane pressure (TMP) culminated at 850 mbar and the flux dropped to 0.5 $L.m^{-2}.hr^{-1}$ LMH until the end of the experiment. The permeate of the SAMBR was low in COD and relatively constant which promoted the growth of autotrophic bacteria in the subsequent AMBR, so that 97.7% of the NH_4^+-N was removed at a maximum nitrogen loading rate of 0.18 kg NH_4^+-N/m^3.day.

Calcium in the leachate was found to precipitate in the AMBR because of the higher pH. A sample taken from the membrane consisted most likely of pure hydroxylapatite $Ca_5(PO_4)_3(OH)$ which had a needle shape, whereas the background of the precipitate consisted more of nodules of calcium carbonate with traces of manganese, iron, magnesium, aluminium, sulphur and sodium.

The very high molecular weight (MW) aromatic organics in the leachate fed to the SAMBRs were almost fully degraded in the bulk of the SAMBRs moreover, their permeate was absolutely free of them which indicates a full rejection of these compounds by the membrane. The medium MW compounds in the range of 395 - 646 kDa were more likely to be rejected by the membrane, while the MW $\leq$ 395 kDa were observed in the permeate. Regarding the evolution of the medium and low MW compounds over the 200 days, it can be stated that overall there was no build-up in the absorbance in any effluent.

1. Introduction

The amount of waste produced in England and Wales, annually, is around 434 million tonnes, of which 8% is Municipal Solid Waste (MSW) (Phillips et al., 2005). Moreover, household waste is growing by around 2% each year. If this rate continues, nearly twice as many new waste management facilities will be required by 2020 to reach UK and EU targets. Municipal Solid Waste (MSW) is a growing concern in the world, and European authorities discourage municipalities choosing landfilling as their waste management strategy. The European landfill directive (99/31/EC) requires that biodegradable municipal waste landfilled should be reduced by 2010 to 75% of that produced in 1995. In addition, the Household Waste Recycling Act 2003 in the UK requires that waste collection authorities shall ensure that by the end of 2010 they should collect at least two types of dry recyclable waste separate from the remainder of the waste. The aim of the Act was to increase the recycling rate of household waste by recycling or composting, and implementation of the Act should allow the UK to reach the target of 25% of MSW being recycled or composted by 2005, 30% by 2010 and 33% by 2015. Unfortunately, in 2003/2004, the majority (72%) of municipal waste in England was still disposed of to landfill, 9% was incinerated and only 19% was recycled or composted (DEFRA, 2005). The last figure increased to 36% in 2008 (DEFRA, 2008), but MSW is still a huge problem.

Currently, biological treatment methods such as composting and Anaerobic Digestion (AD) offer the only route for recycling organic matter and nutrients from the organic fraction of municipal solid waste (OFMSW) (Braber, 1995). Composting can diminish the organic matter by about 50% through the formation of carbon dioxide and water (Held et al., 2002), but it represents an energy consuming process (around 30-35 kWh is consumed per tonne of waste input) while AD is a net energy producing process (100-150 kWh per tonne of input waste). AD is a process by which bacteria degrade organic matter and convert it to mainly carbon dioxide and methane. The main advantage of this technology is the destruction of organic components without the addition of oxygen, and it produces useful by-products such as a gaseous fuel and stabilized solid residue that can be sold as a soil fertilizer. In contrast, aerobic treatment is plagued by the production of large amounts of activated sludge (about 10 times more than AD) and consumption of significant amounts of energy due to aeration requirements (Braber, 1995; Mahmoud, 2002; Speece, 1996). Digester technology for solid waste has basically followed the conventional approach used for sewage sludge digestion over the last fifty years. The result is that the market is currently dominated (over 90%) by single tank designs (Vandevivere et al., 2003) aimed at promoting good mixing. In sewage sludge digestion the rate-limiting factor is the slow growth of the methanogenic population; this necessitates long retention times (15-20 days), which in turn leads to low loadings because of the low solids concentration of the feed sludge. This approach is inappropriate for high solids digestion, however, where the rate-limiting step is in the hydrolysis of the substrate, and hence excessively long retention times are required even at moderate loading rates. In Europe, the first full scale AD plant treating OFMSW were commissioned in the last 1980s and are still in operation demonstrating that it can be considered as a mature technology and more than 168 industrial plants have been built since (Monson et al., 2007).

The anaerobic digestion of solid waste and the organic fraction of municipal solid waste (OFMSW) in particular has been reported to be more stable and robust in two stages (Banks and Humphreys, 1998; Vieitez et al., 2000). The first phase includes a hydrolytic reactor (HR) in which enzymes transform complex molecules into simple ones that can be converted into fatty acids; these acids are then degraded by methanogens in the second reactor. The slow-growing methanogens can therefore operate at an optimum pH independent of the pH in the first reactor. The literature revealed that the hydrolytic stage takes place either in a well-mixed reactor, or in a leach-bed where the leachate is recirculated to the top of the reactor to promote the hydrolysis of solid waste. The loading rate can be as high as 20-25 g $VS.L^{-1}.d^{-1}$ in a Continuously Stirred Tank Reactor depending on the proportion of easily degradable waste, but the hydraulic retention time (HRT) has been restricted to 2 days due to process limitations. In batch leach-bed processes, the OLR is low and they require a large footprint.

In an anaerobic system, the slow growth rate of mixed cultures posed a major problem for reactor design as long minimum solid retention times (SRT) are required. Thus, anaerobic bacteria are easily washed out because their doubling times are normally higher (5-6 days) than the HRT. As efficient anaerobic digester performance depends on the hydrolysis rate of soluble solids and the biomass concentration, the challenge was to develop a reactor which enables the HRT to be uncoupled from the SRT. Thus the objective of this work was to investigate the use of membranes in anaerobic bioreactors to achieve this. Possible problems that may arise include the clogging of the membrane due to the wide molecular weight (MW) distribution from the complex substrates. This may significantly lower the flux which will in turn affect the loading rate to the methanogenic reactor.

A common problem associated with the continuous wet anaerobic fermentation process is that inhibition can occur due to the build-up of light metal ions and ammonia (Gallert et al., 2003). Another issue is the generation of approximately 100-320 L wastewater per ton of waste input (Dierick, 2006; Fricke et al., 2007). Therefore, recycling the stabilized leachate to the head of a continuous wet process treating OFMSW could significantly reduce the amount of wastewater produced, and reduce the environmental impact of MSW disposal. This was one of the objectives of the present work.

Mechanical-biological treatment (MBT) of the OFMSW is now the main strategy to reduce biodegradable MSW in waste in the UK and in Europe (Scaglia and Adani, 2008). It consists of mechanical pretreatment followed by an aerobic (composting-like process) or anaerobic process so that waste impacts are reduced. These processes have attracted attention because they produced a stabilized waste, which has a low impact when disposed of in landfill (Adani et al., 2004). However, landfill leachate still represents a serious environmental threat. In the UK, the leachate from the OFMSW disposed in landfills is currently being treated anaerobically in ponds or aerobically by Sequencing Batch Reactor (SBR) (Robinson, 2005); raw leachate is fed to an aerated reactor. After a certain time, aeration is shut down and the solids are allowed to settle. The supernatant, i.e., the treated leachate, is discharged to sewer or a watercourse. On lab-scale, several researchers have used UASB (Upflow Anaerobic Sludge Blanket), fixed-film and packed bed reactors to treat the leachates from MSW and the COD removal efficiencies were typically higher than 90% (Anderson and Saw, 1992; He et al., 2005). Aerobic Membrane reactors have also been used for biological oxidation and nitrification/ denitrification (Ahn et al., 2002; Laitinen et al., 2006), but the use of an anaerobic MBR for the stabilization of the leachate is not widespread. In particular, Submerged Anaerobic Membrane Bioreactors (SAMBRs) have been developed at Imperial College (Hu, 2004), and stable operation was reported for high strength feed (20,000 mg COD/L) at 20 hours HRT and a permeate flux of 1.5-2 LMH (litres $m^{-2}hour^{-1}$) (Akram, 2006), but no information is available about its behaviour for high-strength leachates from the OFMSW.

The aim of this chapter was thus to explore the feasibility of treating the OFMSW in a two-stage process where the OFMSW is first hydrolyzed in a hydrolytic reactor, and leachate is treated in an anaerobic membrane bioreactor so that the more stable and efficient process can achieve better performance than current technologies. The performance of the process was investigated in a semi-continuous mode, i.e., the HR is fed once every day or every two days. AD of the OFMSW can take place either in dry or wet systems depending on the Total Solids (TS) content of the reactor. For wet fermentation, the dry matter content is adjusted to 8-16% by addition of process water, whereas for dry systems no or only little process water is added to moisten the feedstock. An example of a full scale wet two-stage system is the Schwarting-Uhde process which can sustain an OLR of up to 6 kg VS/m^3.day, whereas a full scale dry 2-stage process such as the BRV plant can achieve up to 8 kg VS/m^3.day (Trosch and Niemann, 1999). When a biomass retention scheme is added such as in the BTA and Biopercolat designs, an OLR up to 15 kg VS/m^3.day can be applied successfully (Gallert et al., 2003; Wellinger et al., 1999). The biofilm growth in the second stage of the Biopercolat process allows the system to run at an overall retention time of 7 days. In the BTA process, the HRT could be reduced to 5.7 days.

For laboratory and pilot scale anaerobic leachate treatment experiences, OLRs from 3 to 22 kg COD/m^3.day with COD removal efficiencies of 68 - 97% and HRTs between 1.5 and 2.6 days have been reported previously (Chang, 1989; Henry et al., 1987; Kennedy et al., 1988). On the other hand, aerobic leachate treatments found in the literature have been applied to leachate with COD values between 3,000 and 48,000 mg/L. COD removal efficiencies reported for aerobic systems are higher than 70%, with HRTs ranging from 2.5 to 20 days (Boyle and Ham, 1974; Cook and Foree, 1974; Maris et al., 1984; Robinson and Maris, 1985; Uloth and Mavinic, 1977). However, less sludge is generated and less energy is required if an anaerobic step is followed by an aerobic one. In this process sequence, the final aerobic stage serves as a post-treatment to improve the final effluent quality (Agdag and Sponza, 2005; Hoilijoki et al., 2000). For instance, Borzacconi et al. (1999) loaded a UASB at an OLR of 20 kg COD/m^3.day at an HRT of 2 days, and achieved a COD removal greater than 80%; the subsequent aerobic rotating biological contactor achieved 72% COD removal. Xu et al. (2008) removed 7% of the SCOD in an anaerobic filter, while 82% was removed in the subsequent aerobic MBR at 9.5 days HRT. They pointed out that most of the polycyclic aromatic hydrocarbons (PAHs) were biodegraded in the anaerobic step. The final effluent, however, still contained 1,000 mg SCOD/L which suggests that there were some biorefractory organics that were hardly removed by both biological processes and the membrane process. Another process advantage is the possibility of removing ammonia from leachate in the aerobic step, but it is known that high influent COD promotes heterotrophic growth and inhibits ammonium oxidation (Cheng and Chen, 1994; Hanaki et al., 1990). Different process configurations have been reported for the simultaneous removal of COD and ammonia from landfill leachate. Im et al. (2001) used an up-flow anaerobic biofilm reactor (36°C), an aerobic activated sludge reactor (23°C) and a clarifier achieving an organic removal rate of 15.2 kg COD/m^3.d in the anaerobic reactor and ammonium removal rate of 0.84 kg N/m^3.day in the aerobic reactor operating at 4 days HRT. Agdag and Sponza (2005) obtained 98% COD removal of food waste leachate at an OLR of 16 kg COD/m^3.d in two UASBs (HRT=1.25 day) and an aerobic CSTR used in sequence. Ninety nine percent of NH_4^+ was removed at 4.5 days HRT in the aerobic CSTR. Chen et al. (2008) used an anaerobic-aerobic moving-bed biofilm system and achieved a COD removal of 92% at an OLR of 15.7 kg COD/m^3.d, while 97% of NH_4^+-N was removed when the HRT of the aerobic step was more than 1.25 days. Jokela et al. (2002) obtained over 90% nitrification at 0.13 kg N/m^3.day at 25°C and 1.4 day HRT in an upflow filter with crushed bricks.

The objectives of this chapter were numerous: the effect of the inoculum on the behaviour of the SAMBR was investigated; the stability of the SAMBRs and the AMBR was tested at different HRTs and OLRs; and an AMBR operating at ambient temperature was set up to determine whether the recalcitrants from the SAMBR could be biodegraded aerobically. After 200 days of operation, another objective was to see if there was a build-up of recalcitrants with time due to the permeate recycle, or if there was slow degradation, and GC-MS analysis was performed to determine what if any these recalcitrants were. Another pertinent question related to continuous wet anaerobic fermentation process with effluent recycle is whether light metals ions (Na^+, K^+, Mg^{2+}, Ca^{2+}) and Cl^-, PO_4^{3-}, SO_4^{2-} and ammonia accumulate to inhibitory levels (Gallert et al., 2003) and these parameters were therefore monitored throughout the 200 days of operation. Leachate recirculation over a tank filled with MSW is relatively well documented (Bilgili et al., 2007; Hao et al., 2008), but recirculation of stabilized leachate from membrane bioreactors is not. During anaerobic treatment of organic wastes, approximately 100-170 L wastewater per ton of waste input are generated (Fricke et al., 2007). Therefore, recycling the stabilized leachate to the head of a continuous wet process treating OFMSW could significantly reduce the amount of wastewater produced, and reduce the environmental impact of MSW disposal.

2. Materials and Methods

2.1. Experimental Setup

2.1.1. Hydrolytic Reactor

Figure 1. 3-D view of the hydrolytic reactor (left) and stirrer (right).

The hydrolytic reactor (HR) is a 20 liter cylinder made of acrylic sheet; a stainless steel mesh follows a concentric arrangement inside the cylinder and has a grid of 1mm square holes. A stirrer moves inside the mesh allowing two pieces of rubber to rub against it. Figure 1 shows the reactor and the stirrer. The reactor is wrapped in a copper coil in which hot water is circulated and glass fiber is used to cover the coil so that the reactor internal temperature is maintained at 35 ± 1°C. The stirrer speed is set at 64 rpm.

The HR (10L working volume) was operated intermittently (15 min ON-15 min OFF). The HR was fitted with a 50 micron stainless steel macro filter (Spectrum Laboratories Inc.) on the inside of the stainless steel mesh (1 mm) in order to retain the large partially hydrolyzed particles and thereby separate the coarse solids from the leachate being fed to the SAMBRs, so that the SRT and the HRT were uncoupled.

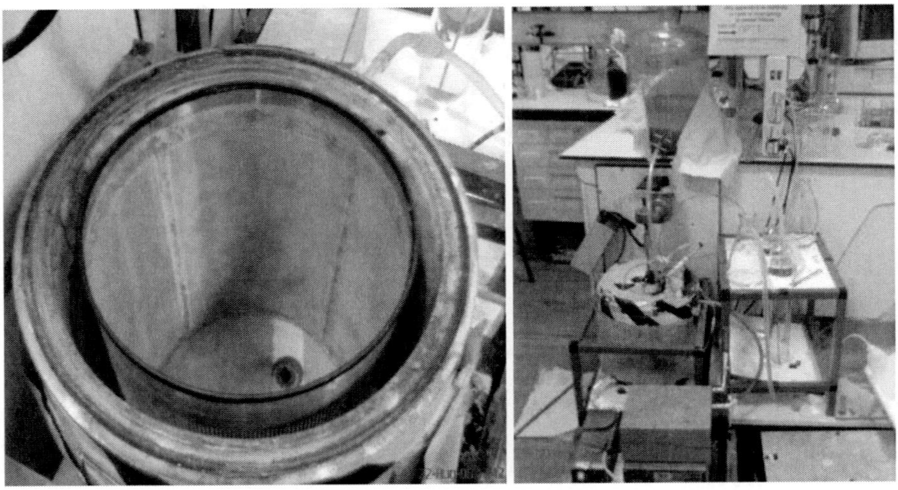

Figure 2. Inside view of the Hydrolytic Reactor (HR) without the stirrer (left). Photo of the laboratory HR set-up (Right).

The HR was inoculated with 4L of biomass from a previous batch test in the HR at a residence time of about 50 days. The inoculum was sieved through a 180 micron screen and its TSS and VSS were 2.74 and 2.07 g/L, respectively. The HR was loaded with 400 g OFMSW on a dry matter basis (≈ 340 g VS), and the volume was adjusted to 10L with tap water containing $NaHCO_3$ so that the HR was started up at 4,000 mg equivalent $CaCO_3$/L of alkalinity. The HR was then fed semi-continuously with a feedstock of 10% Total Solids that was prepared by adding leachate from the HR to the simulated OFMSW in order to blend the mixture and obtain a homogeneous slurry. Fresh tap water was only added to the HR to keep a constant working volume. Until day 159, the HR was fed once every two days, then from day 160 onwards, it was fed once a day. The digestate from the HR was removed just before feeding, and it was dewatered by hand compression over a 500 microns sieve. The liquid fraction was used to moisten the fresh feedstock, while a fraction of the dewatered digestate was taken, weighed and discarded to set a specific SRT in the HR.

2.1.2. Methanogenic Reactor

A schematic diagram of the experimental set up used is shown in Figure 3. Each submerged anaerobic membrane bioreactor (SAMBR) has a working volume of 3 litres, and is made of acrylic panels. Each reactor contains a standing baffle designed to direct the fluid to the upcomer and downcomer regimes. The reactors were maintained at 35 ± 1°C, and the biomass was continuously mixed using headspace biogas that was pumped (Charles Austen Pumps, Model B100SEC) through a stainless steel tube diffuser to generate coarse bubbles. The bubbles push the sludge flow upward between the membrane module and the reactor wall in the upper section. A flask (500 mL) was placed before the biogas pump in order to collect any sludge going to the gas line due to excessive foaming or excess of liquid in the reactor so that the biogas pump was protected against any liquid that could harm the mechanism. The sparging rate was controlled by a gas flowmeter (2 - 20 LPM, ColeParmer, USA) to minimize membrane fouling. A data logging system (USB-1408FS, Measurement computing) was used to monitor transmembrane pressure (PMP1400, 1 bar A, Druck), and permeate flow rates (13 – 100 mL/min, Flow Sensor Ryton Sensor, McMillan). The membrane module (Kubota) had 0.1 m^2 of total surface area with a pore size of 0.4 micron. The membrane panel was comprised of a solid acrylonitrile butadiene styrene support plate (5 mm thick) with a spacer layer between it, and a welded flat sheet membrane, which was made of polyethylene, on both side. The membrane was coated with a special chemical (10%w ethanol) to give a hydrophilic surface to the membrane.

The HR and SAMBR1 were connected in series: the leachate was fed to SAMBR1 and the permeate from SAMBR1 was recycled to the HR in order to maintain the moisture and alkalinity of the system. On day 45, SAMBR2 was fed on leachate in parallel with SAMBR1 in order to compare the effect of inoculum on the start-up of SAMBR. The HR, SAMBR1 and SAMBR2 were maintained at 35 ± 1°C. The biogas sparging rate was set at 5 L/min (LPM) to minimize cake formation on the membrane. On day 130, an Aerobic Membrane Bioreactor (AMBR) operating at ambient temperature (21-22°C) was started up to treat the permeate of SAMBR2. The permeate of the AMBR was then returned to the HR. The two SAMBRs and the AMBR were three litre reactors fitted with a Kubota polyethylene flat sheet membrane of 0.1 m^2 of total surface and a pore size of 0.4 microns.

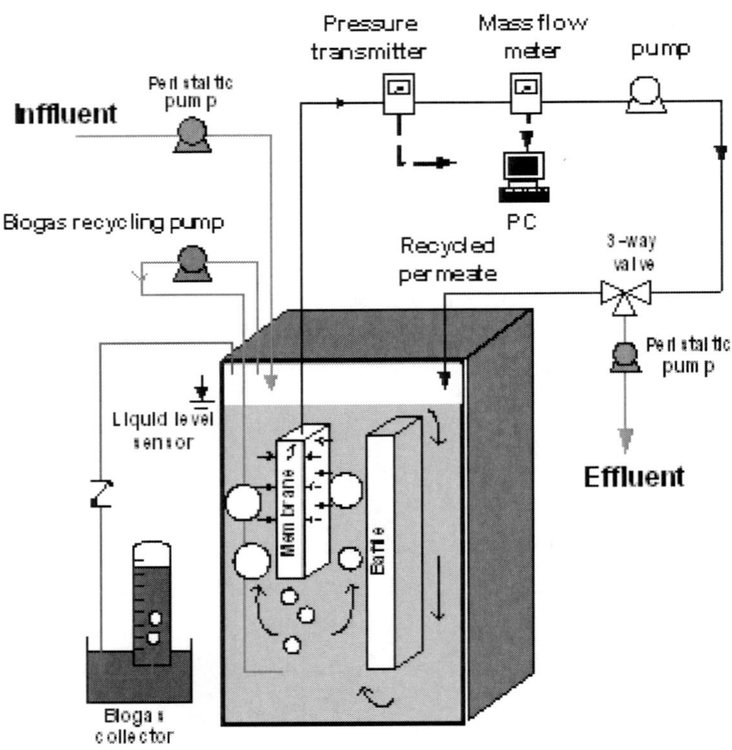

Figure 3. Schematic diagram of the Submerged Anaerobic Membrane Bioreactor.

SAMBR1 was inoculated with 0.5 L of seed from a SAMBR fed on leachate from the same simulated OFMSW at a HRT of 4 days (Figure 4). The volume was adjusted to 3 L with the anaerobic biomedium defined in

Owen et al. (1979) so that the initial TSS and VSS were 3.31 and 2.54 g/L, respectively.

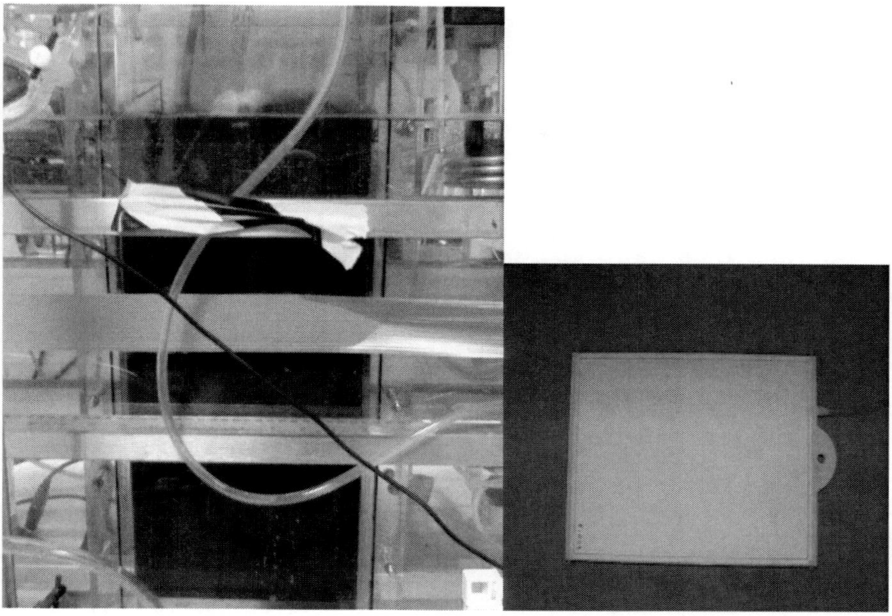

Figure 4. Left: photo of the Submerged Anaerobic Membrane Bioreactor (SAMBR) during start-up. The tip of the submerged membrane module can be seen above the liquid. Right:Kubota membrane module in the SAMBR.

SAMBR2 was inoculated with biomass from a 4 litre chemostat batch-fed (once a week) on a 8 g COD/L feed with a composition given elsewhere (Nachaiyasit and Stuckey, 1995). The feed consisted of peptone and meat extract (25% on a COD basis) and a synthetic VFAs mixture (75% on a COD basis). The ratios of the VFAs compared to acetic acid were 1.2, 0.05, 0.22, 0.08, 0.23 for propionate, iso-butyrate, n-butyrate, iso-valerate, n-valerate, respectively. These ratios were typically observed in the raw leachate obtained in previous tests from the simulated OFMSW. The supernatant of the chemostat was discarded and the settled solids were used to inoculate SAMBR2. The volume was adjusted to 3L with the anaerobic biomedium defined in Owen et al. (1979) so that the initial TSS and VSS were 2.56 and 1.78 g/L, respectively.

The AMBR was inoculated with biomass from a dye wastewater plant at an initial MLTSS and MLVSS of 3 and 2.3 g/l, respectively. Air was used to mix the reactor content at 2 LPM.

2.2. Analytical Methods

2.2.1. pH

pH was measured using a calibrated pH meter (Jenway, Model 3020). Values obtained were accurate to within ±0.02 units.

2.2.2. Oxido-Reduction Potential

The oxido-reduction potential (ORP) was measured with a platinum-band ORP electrode (Cole-Palmer, USA) connected to a Jenway 3020 pH meter. Values obtained were accurate to within ± 3 mV.

2.2.3. Biogas Composition and Production Rate

The composition of biogas was determined using a Shimadzu GC-TCD fitted with a Porapak N column (1500 × 6.35 mm). The carrier gas was Helium set at a flow rate of 50 ml/min. The column, detector and injector temperature were 28, 38 and 128°C, respectively. The peak areas were calculated and printed out on a Shimazdu Chromatopac C-R6A integrator. Samples of 1 ml were collected using 1 ml plastic syringes (Terumo). The coefficient of variation for 10 identical samples was ± 2%.

In the HR, the gas production rate was measured using Tedlar bags connected on the top. To empty the Tedlar bag, it was connected to the top of a column full of water. A valve at the bottom of the column was then opened to let the water drain into a collection vessel as the gas filled the column. The water stopped running when the Tedlar bag was empty which enabled the volume of biogas produced to be measured. The percentage of methane was analyzed by taking a sample at the top of the column. In the SAMBR, the gas production rate was measured using the water displacement method with cylinders filled with water acidified with 2%

(v/v) H_2SO_4 and 10% (w/v) NaCl according to Agdag and Sponza (2007). The top of the SAMBR was connected to the top of the cylinders, and when gas was produced the water left the cylinder from the bottom and was collected in a vessel placed above it. A one-way valve was placed between the cylinder and the collection vessel so that the water in the collection vessel did not exert any pressure on the gas in the cylinder.

2.2.4. Volatile Fatty Acids (VFAs)

For acetic, propionic, n-butyric, iso-butyric, n-valeric, iso-valeric and n-caproic acids analysis, the sample was filtered through a 0.2 micron filter, and then acidified by adding one drop of concentrated sulphuric acid, and eventually analyzed using a Shimadzu Gas Chromatograph with a flame-ionised detector and a SGE capillary column (12m x 0.53mm ID-BP21 0.5micron). The mobile phase was helium and the injector, column and detector temperature were 200, 80 and 250°C, respectively. The relative standard deviation was ± 3% for ten identical samples.

2.2.5. Gas Chromatography - Mass Spectrometry (GC-MS)

For the GC-MS analysis, the analytes of interest (low molecular weight hydrophobic compounds) were extracted using a solid phase extraction (SPE) procedure. The Oasis HLB cartridge (Waters Corporation) was first conditioned with 3 mL methyl tertiary-butyl ether (MTBE), 3mL methanol and 3 mL deionized water (DW). A sample (500mL) at pH 2 was then loaded onto the cartridge and filtered dropwise. The cartridge was then washed with 3mL of 40% methanol in DW to remove organic interferences, re-equilibrated with 3mL DW, washed with 3mL 10% methanol/2% NH_4OH to remove humic interferences and finally 6mL 10% methanol/90% MTBE. The final matrix was then evaporated to 200 microliter.

The samples were then analyzed using a 5890 Series gas chromatograph equipped with an autosampler and a 5970 mass spectrometry detector (Hewlett-Packard, USA). Analytes were separated using a SGE HT5 column of 25m x 0.22mm with a film thickness of 0.1

micron. The temperature program was: 50°C, hold 2 min, rate 8°C min-1 to 350°C, hold 30 seconds. Helium was used as a carrier gas at a flowrate of 2 ml/min. The injector temperature was set at 270°C. The MS was operated in the electron impact ionisation mode (70eV). The transfer line and ion source temperatures were 290°C and 220°C, respectively, and the quadrupole was not heated. Scan runs were made with a range from m/z 33 to 500. The chromatograms were analysed using the NIST05 library and the compound was deemed identified if the match percentage was higher than 70%. A retention index (RI) was attributed to the unidentified peaks according to Van den Dool and Kratz (Van Den Doole and Kratz, 1963):

$$RI = 100\,.\left[z + \frac{t_i - t_z}{t_{z+1} - t_z}\right]$$

where t_z and t_{z+1} are retention times of the reference n-alkanes hydrocarbons eluting immediately before and after the chemical compound i. Table 1 lists the retention times of the n-alkane hydrocarbon standard used in the Van den Doole-Kartz retention index equation. Regarding the concentrations of the analytes, two sets of standards were run with o-hydroxybiphenyl and bis(2-ethylhexyl)phthalate which are components detected in this study. Standards of 20, 2 and 0.2 mg/L were used to obtain the calibration with a coefficient of correlation R^2 greater than 0.99 in both cases. The limit of detection of the GC-MS used in this study was 50 ng/L.

2.2.6. Particle Size Distribution

Particle size measurements were made using a Malvern Instruments Particle Size Analyser Model 2600C with a helium neon laser. The standard deviation was within ± 5%. The value reported for this parameter is the d(4,3) value, which is the volume mean diameter of particles, which takes into account the number and volume of particles in a particular medium. This is different from the Sauter mean d(3,2) which can be defined as the diameter of a sphere that has the same volume/surface area ratio as a particle of interest.

2.2.7. Heavy Metals

All the glassware was washed with Aqua Regia (1 volume of concentrated HNO_3 with 3 volumes of concentrated HCl) prior to the experiment. For the determination of heavy metals in a solid sample, Aqua Regia was used to digest a pre-weighted sample.

Table 1. Retention times of standard hydrocarbons detected by the GC-MS used in this study

Retention time (min)	Hydrocarbons	Formula
6.568	decane	$C_{10}H_{22}$
7.256	undecane	$C_{11}H_{24}$
9.219	dodecane	$C_{12}H_{26}$
11.083	tridecane	$C_{13}H_{28}$
12.866	tetradecane	$C_{14}H_{30}$
14.532	pentadecane	$C_{15}H_{32}$
16.109	hexadecane	$C_{16}H_{34}$
17.614	heptadecane	$C_{17}H_{36}$
19.014	octadecane	$C_{18}H_{38}$
20.44	nonadecane	$C_{19}H_{40}$
21.737	eicosane	$C_{20}H_{42}$
23.028	heneicosane	$C_{21}H_{44}$
24.16	docosane	$C_{22}H_{46}$
25.194	tricosane	$C_{23}H_{48}$
26.338	tetracosane	$C_{24}H_{50}$
27.337	pentacosane	$C_{25}H_{52}$
28.355	hexacosane	$C_{26}H_{54}$
29.33	heptacosane	$C_{27}H_{56}$
30.267	octacosane	$C_{28}H_{58}$
31.174	nonacosane	$C_{29}H_{60}$
32.074	triacontane	$C_{30}H_{62}$

The solution was heated for 1 hour with a watch glass and was not allowed to boil. After cooling, the watch glass and the beaker were rinsed with distilled water and the final solution was filtered through a 0.45 micron filter.

A blank with deionized water was run following the same procedure. The sample and blank were analyzed with PerkinElmer Optima 2000 DV Inductively Coupled Plasma-Optical Emission Spectrometer (ICP-OES) in triplicate. The coefficient of variation is different for each element and is given along with the results. The detection limit is given in Table 2 for each element.

Table 2. Detection limits of heavy metals analyzed by Inductively Coupled Plasma-Optical Emission Spectrometer

Element	Detection Limit (ppb)
Al	10
B	10
Cd	1
Co	2
Cr	2
Cu	4
Fe	1
Mn	1
Mo	5
Ni	5
Pb	10
V	5
W	10
Zn	2
Na	5
K	5
Ca	0.5
Mg	1

2.2.8. Ion Chromatography

The ions Na^+, K^+, Mg^{++}, Ca^{2+}, Cl^-, NO_2^-, NO_3^-, PO_4^{3-}, SO_4^{2-} were analysed using the DIONEX Ion Chromatograph as described by the American Public Health Association (APHA, 1999). The coefficient of variation for the ions listed above are respectively 2, 0.2, 0.5, 0.6, 1.1, 1.8, 1.8, 3.4 and 5.1 for 5 identical samples. The detection limit was between 50 and 100 ppb.

2.2.9. Size Exclusion Chromatography (SEC)

For size exclusion chromatography (SEC), an Aquagel OH-40 column (Polymer Labs) was used with deionised water as eluent delivered at a flowrate of 0.5 mL/min. The sample volume was 50 microliter, and the column was maintained at ambient temperature with a Shimadzu UV detector set at 254 nm. Unbranched standards of polyethylene oxide (PEO) and polyethylene glycol (PEG) and glucose were used to calibrate the system, hence the results obtained are quoted relative to these linear standards detected by a Shimadzu refractive index detector.

2.2.10. Scanning Electron Microscope - Energy Dispersive X-ray (SEM-EDX)

Fouled membranes and sludge samples were analyzed by scanning electron microscopy (SEM) and energy dispersive X-ray (EDX) spectroscopy. SEM samples were fixed overnight at 4°C in 3% glutaraldehyde and kept at pH 7.2 by a 0.1M phosphate buffer. Samples were then dehydrated in a graded ethanol/water series (10-30-50-70-90-100%) for 20 minutes at each concentration, and then dried for a day at 30°C. Samples were sputtered-coated with gold or carbon (30mA for 2.5 minutes, vacuum 0.2 Torr) prior to SEM/EDX analyses. Specimens were examined and photographed under a scanning electron microscopy (JEOL JSM-5610LV). Energy Dispersive X-Ray analysis was conducted with an EDX-60 (Oxford instrument-incax-sight). The EDX analyzer was connected to a scanning electron microscope (model JSM-840A). More than one point on the sample was analyzed by EDX. The exact number of replicates is given along with the results.

2.3. Physico-Chemical Methods

2.3.1. Alkalinity

The alkalinity of wastewater is based on its acid-neutralizing capacity, and is the sum of all the titratable bases. The measured value may vary significantly with the end-point pH used. The method used in this chapter followed the procedure described in Standard Methods (APHA, 1999). Alkalinity of samples were titrated potentiometrically to an end-point of pH 4.5 with 0.1 N H_2SO_4. The sulphuric acid solution was prepared by dissolving 2.8 mL of concentrated sulphuric acid in 1L of deionised water. The coefficient of variation for 10 identical samples was within 2.7%. The calculation of alkalinity is shown below:

$$\text{Alkalinity (as mg } CaCO_3\text{/L)} = \frac{A \; X \; N \; X \; 50{,}000}{mL \; sample} \quad (1)$$

where: A= mL standard acid used, N = normality of standard acid.

2.3.2. Total Solids, Volatile Solids and Fixed Solids

The measurement of Total Solids (TS) and Volatile Solids (VS) was applied as described in Standard Methods (APHA, 1999). Empty aluminium dishes were placed in a furnace at 550°C for 1 hour. After cooling in a dessicator, the sample was injected into the pre-weighted dish, and the liquid fraction evaporated overnight in an oven at 103-105°C until the weight stabilized. The resulting weight was recorded for the measurement of TS and then the dish was placed in a furnace at 550°C for 1 hour. For VS measurement, the final weight was then recorded on removal from the furnace. Calculations were then performed as given in Standard Methods (APHA, 1999). Fixed solids (FS) or ash percentage was simply obtained by difference. The coefficient of variation for 10 identical samples was within 4%.

2.3.3. Total Suspended Solids and Volatile Suspended Solids

The measurement of Total Suspended Solids (TSS) and Volatile Suspended Solids (VSS) was carried out according to Sections 2540-B and 2540-E of Standard Methods (APHA, 1999). Firstly, three glass fiber filters (GFC, Whatman) were washed by filtering 20 mL deionized 3 times. The filters were placed on aluminium dishes and placed in a furnace at 550°C for 1 hour. They were then placed in a dessicator until needed. To measure TSS, a known volume of sludge was filtered through the filter, and the liquid fraction was evaporated overnight in an oven at 103-105°C. They were then allowed to cool down in a dessicator. The resulting weight was recorded for the measurement of TSS and then the tray was placed in a furnace at 550°C for 1 hour. For VSS measurement, the final weight was then recorded on removal from the furnace. Fixed Suspended Solids was obtained by subtracting VSS from TSS. Calculations were then performed as given in Standard Methods (APHA, 1999). The coefficient of variation (COV) for ten identical samples was 4%, 3.1% and 7.1% for TSS, VSS and FSS, respectively.

2.3.4. COD Measurement

The measurement of Chemical Oxygen Demand (COD) was based on the Standard Closed Reflux Colorimetric Method described in Section 5220-D of Standard Methods (APHA, 1999). Digestion solution was first prepared by adding 10.216 g of $K_2Cr_2O_7$ (Merck), previously dried overnight at 103°C, 167 ml of concentrated H_2SO_4 (Merck, UK) and 33.3 g of $HgSO_4$ (Merck, UK) into 500 ml of distilled water. The mixture was then left to cool at room temperature before diluting to 1000 ml. Samples of 1 ml were added to a Hach reflux tube, followed by 0.6 ml of digestion solution. Then 1.4 ml of sulphuric acid reagent (2.5% w/w silver sulphate in sulphuric acid) was carefully run down the inside of the tube so that an acid layer was formed under the sample/digestion solution layer. The tubes were tightly sealed and inverted three times to mix properly. The mixtures were then refluxed in a Hach COD reflux reactor (Model 45600) at 150°C for 2 hours. After cooling, the samples were analysed on a Shimadzu UV/VIS scanning spectrophotometer (Model UV-2101/3101 PC) at a wavelength of 600 nm. Potassium hydrogen phthalate (KHP) (Merck, UK) was used to prepare standard solutions in the range of 20-900 mg/l. KHP has a theoretical COD of 1.176 mg O_2/mg.

Soluble COD (SCOD) was obtained by centrifuging the samples at 10,000 rpm for 5 minutes (Biofuge Stratos, Heraeus Instruments), then at 17,000 rpm for 20 minutes to ensure an easy filtration through a 0.2 micron filter to remove fine suspended material and any residual biomass. The coefficient of variance for ten identical samples was ± 2.6%.

Total COD (TCOD) was obtained by taking 1ml of sample and diluting it to 100 ml with deionised water. Although the sample was blended for 1 minute in a 1L Waring blender, the presence of particles in suspension made it difficult to take representative samples. The coefficient of variance for ten identical samples was 9.9%.

2.3.5. Ammonia Measurement

The measurement of ammonia was carried out according to the Nesslerization method in Standard Methods (APHA, 1999). A sample of 0.25 ml was diluted with deionized water to 25 mL in a volumetric flask. Three drops of mineral stabilizer (Hach Lange GmbH) were added followed by three drops of polyvinyl alcohol dispersing agent (Hach Lange GmbH). The mixture was then mixed by inverting the flask. One milliliter of Nessler reagent (VWR) was then added and the yellow colour was then analyzed after ten minutes on the spectrophotometer at a wavelength of 495nm. The coefficient of variance for ten identical samples was ± 6.6%.

2.3.6. Total Nitrogen Measurement

The measurement of total Nitrogen was carried out according to the TN kit manual from Hach Lange GmbH. An alkaline persulfate digestion converted all forms of nitrogen to nitrate. Sodium metabisulfite was added after the digestion to eliminate halogen oxide interferences. Nitrate then reacts with chromotropic acid under strongly acidic conditions to form a yellow complex with an absorbance maximum at 410 nm. The coefficient of variance for ten identical samples was ± 5.5%.

2.3.7. Total Phosphorus Measurement

For the measurement of Total Phosphorus (TP) a known mass of sample was digested in a mixture of hydrochloric and nitric acids (3:1) at 150°C until no solid particles were visible. The digested acid mixture was then transferred to a 50 mL flask, 1 drop of phenolphtalein indicator solution was added and sodium hydroxyde was added until a faint pink color appeared. Deionized water was added to reach the 50 ml mark. The measurement of TP was then carried out according to the vanadomolybdophosphoric acid colorimetric method described in Standard Methods (APHA, 1999): in a dilute orthophosphate solution, ammonium molybdate reacts under acid conditions to form a heteropoly acid, molybdophosphoric acid. In the presence of vanadium, yellow vanadomolybdophosphoric acid is formed. The intensity of the yellow colour is proportional to phosphate concentration. The absorbance was read on a spectrophotometer at 470 nm. The coefficient of variance for ten identical samples was ± 0.6%.

2.3.8. Carbohydrates Measurement

Carbohydrates are hydrolysed by sulphuric acid; dextrose and phenol, in the presence of sulphuric acid, react further to produce a colored complex that is measured spectrophotometrically. Samples of 0.4 ml were added to each test tube and then mixed well with 0.4 ml of 5% (w/v) phenol (Sigma, UK). Two ml of H_2SO_4 was added to each test tube and left at room temperature (18-26°C) for 20-30 minutes to allow color development. The content of each tube was transferred to cuvettes and the samples were analysed on a Shimadzu UV/VIS scanning spectrophotometer (Model UV-2101/3101 PC) against the blank at a wavelength of 485 nm (Dubois et al., 1956). Dextrose was used to prepare standard solutions in the range of 0 - 200 mg/l, but the method gives a linear relationship for some five carbon sugars such as D-xylose and Darabinose. To convert into COD, 1g carbohydrates assumed as $C_6H_{12}O_6$ is equivalent 1.07 g COD (Sanders, 2001). The coefficient of variance was within 2.5% for ten identical samples.

2.3.9. Protein Measurement

A Protein kit (Stephenson's modification) was purchased from Sigma, England (Kit no. P5656). Samples of 1 ml were added to each tube with 1 ml of Lowry Reagent Solution. The samples were mixed well and kept at room temperature for 20 minutes. With rapid and immediate mixing, 0.5 ml of Folin and Ciocalteu's Phenol reagent was added to each tube and mixed well after addition. These samples were left at room temperature for 30 minutes. The content of each tube was transferred to cuvettes and the samples were analysed on a Shimadzu UV/VIS scanning spectrophotometer (Model UV-2101/3101 PC) against the blank at a wavelength of 725 nm. The detection limit was 5 mg/L, and the coefficient of variance within 3.6% for 10 identical samples. As the precise chemical formula of the proteins detected was not determined, the percentage of soluble COD represented by protein had to be estimated by assuming a stoichiometric conversion factor of 1.5 which is derived from the typical formula of proteins ($C_{16}H_{24}O_5N_4$) presented in Rittmann and McCarty (Rittman and McCarty, 2001).

2.3.10. Biochemical Methane Potential Test

The assay was conducted using the media and serum bottle techniques developed by Owen et al. (Owen et al., 1979). The gas used to purge the assay bottles during preparation of the assays was a mixture of 70% N_2 and 30% CO_2 at a flow rate of approximately 0.5 L.min-1. Two mL sample, 14 mL biomedia and 4 mL biomass were transferred into 20-ml serum bottles under anaerobic conditions by continuously flushing the bottles with a mixture of N_2 and CO_2. Blanks were run with no carbon source and the methane obtained in these control bottles was due to the autolysis of the inoculum. Each serum bottle was then capped with leak proof Teflon seals. Ultimate Methane potential was measured by the amount of methane produced (converted to standard temperature and pressure) minus the methane produced by the blanks and divided by the mass of volatile solids added. Gas production was measured using a glass syringe and the gas was then wasted, whereas gas composition was measured with the GC described above. Duplicate samples were run and the coefficient of variation in the worst case was ± 6%.

2.4. Molecular Biology Methods: Denaturing Gel Gradient Electrophoresis (DGGE)

The DNA from a mixed culture was extracted using the FastDNA SPin for soil kit from MP Biomedicals. The PCR mixture (50 µL) contained 2 µL of each primer, 0.2 µL of the Taq Polymerase (Promega), 2 µL of dNTPs stock solution (10 µM), 5 µL of Taq buffer solution containing $MgCl_2$ and between 1 and 20 µL of DNA template. The final volume was adjusted to 50 µL using DNA-free water. Archaeal DNA was amplified using a nested PCR reaction in a G-Storm thermocycler. First, the DNA template was amplified using the primers 46F and 1017R in 25µL PCR volume. The temperature program was: initial denaturation 95°C for 3 minutes, then 35 cycles of denaturation at 95°C for 1 minute, annealing at 40°C for 1 minute, elongation at 72°C for 1 minute and the final elongation took place at 72°C for 7 minutes (Akarsubasi et al., 2005; Gray et al., 2002). Then, 1 microliter of the PCR product was used for the second PCR reaction in 50 µL using the primers 344F-GC and Univ522R. The temperature program was: initial denaturation 95°C for 3 minutes, then 35 cycles of 95°C for 1 minute, 53°C for 1 minute, 72°C for 1 minute and the final elongation took place at 72°C for 7 minutes (Akarsubasi et al., 2005; Gray et al., 2002).

Bacterial DNA was amplified using a single PCR reaction in the same thermocycler. The bacterial DNA was amplified using the primers 341F-GC and 907R (Fernandez et al., 2008; Liu et al., 2008) in 50 µL PCR volume. The temperature program was: initial denaturation 94°C for 5 minutes, then 30 cycles of 94°C for 1 minute, 52°C for 1 minute, 72°C for 1 minute and the final elongation took place at 72°C for 10 minutes. The DNA specific to ammonia-oxidisers was amplified using the primers CTO189FGC and CTO654R in 50µL PCR volume. The temperature program was: initial denaturation 94°C for 1 minutes, then 35 cycles of 92°C for 30 seconds, 57°C for 1 minute, 68°C for 45 seconds and the final elongation took place at 68°C for 5 minutes (Kowalchuk et al., 1998; Stephen et al., 1998).

The DNA specific to fungi was amplified using the primers ITS4 and ITS5 in 50µL PCR volume. The temperature program was: initial denaturation 95°C for 1 minutes, then 30 cycles of 95°C for 1 minute, 58°C for 30 seconds, 72°C for 1 minute and the final elongation took place at 72°C for 10 minutes (White et al., 1990). Presence of PCR products was confirmed by electrophoresis on 1% agarose gels stained with ethidium bromide. DGGE of the 16Sr DNA PCR products was carried out using the DCODETM system (Bio-Rad Laboratories Ltd) according to the manufacturer's instructions and protocols. The percentage of polyacrylamide varied according to the range of base pairs (Table 3) as recommended in the manufacturer's manual:

Table 3. Polyacrylamide concentration in the DGGE vs. the base pair of the PCR products

Gel %	**base pair**
6%	300-1000 bp
8%	200-400 bp
10%	100-300 bp

This implied that a 6% gel was made up if bacterial primers were used, while a 8% or even a 10% gel was required for archaeal primers. The PCR products were electrophorised at 60V for 16 hours at 65°C with a denaturing gradient ranging from 40 to 60% for bacterial primers (Liu et al., 2008), 45-65% for archeal primers and 38-50% for ammonia-oxidising bacteria (Kowalchuk et al., 1998; Stephen et al., 1998). A 100% solution contains 7 mol/L urea and 40% formamide. The gel was stained in 1 X TAE buffer containing SYBR green before visualizing on a UV transilluminator and photographed. The brightest DGGE bands were cut out, eluted in DNA-free water overnight at 4°C and then re-amplified. The purified PCR product was then cloned with the pCR 2.1-TOPO cloning kit (Invitrogen) according to the manual instructions. For the cloning reaction, 1 μL of vector was mixed carefully with 1 μL of salt solution (Invitrogen) and 4 μL of purified PCR product. Meanwhile chemically competent *E. coli* were placed on ice and 2 μL of the cloning reaction was mixed with the competent cells. The cells were transformed using a heat shock protocol (30 seconds at 42°C), then incubated for 1 hour at 37°C in S.O.C. medium and eventually spread onto Luria-Bertani (LB) agar plates containing 50 μg/mL of Kanamycin to incubate overnight at 37°C. The LB medium consisted of tryptone (10g/L), yeast extract (5g/L), NaCl (10g/L) and 15g/L of agar in 1L deionized water. The pH was adjusted to 7 and the medium was autoclaved (120°C for 20 min.). Kanamycin was added in a sterile environment. Three well-separated colonies were picked from the plate using a pipette tip and were grown between 12 and 16 hours in LB medium (without agar) at 37°C. The plasmid DNA was then purified using the purification kit from Qiagen (QIAprep Spin Miniprep Kit) and the resulting DNA was then kept frozen until it was sent off for sequencing (Cogenics, UK). DNA sequences analyses were performed using the BLAST server of the National Center for Biotechnology Information (http://www.ncbi.nlm.nih.gov).

The Expect value (E) is a parameter that describes the number of hits one can "expect" to see by chance when searching a database of a particular size. It decreases exponentially as the percentage of the match increases. Essentially, the E value describes the random background noise. For example, an E value of 1 assigned to a hit can be interpreted as meaning that in a database of the current size one might expect to see 1 match with a similar score simply by chance. The lower the E-value, or the closer it is to zero, the more "significant" the match is.

2.5. Feedstock Composition and Properties

2.5.1. Composition

The simulated OFMSW used for this study was made up of paper, food wastes and garden wastes. Kitchen wastes (KW) came from a canteen in Southampton University; the left-overs were passed through a kitchen grinder and mixed in a large tank with a drill mixer and then frozen until the experiment. The composition of the simulated paper waste used for the study is listed in Table 4. Paper wastes were kept at room temperature. Garden waste was collected from the Downend Quarry centralised composting site near Fareham (Hampshire, UK) and were kept at 4°C until the experiment.

The OFMSW feedstock used in this study consisted of 41.3% Kitchen Wastes (KW), 10.8% Garden Wastes (GW) and 47.9% Paper Wastes (PW) on a wet basis. This composition comes from an assessment scheme for kerbside collection of dry recyclables undertaken by the district of Eastleigh in the UK.

Table 4. Composition of the simulated paper waste used for the study

Type of paper	**%**
Newspaper	21.2
Magazine	12
Office paper	7.9
Card and paper packaging	10.5
Cardboard	1.2
Card non packaging	0.6
Liquid carton	1.4
Tissue paper	15.06
Paper plate	15.06
Toilet paper	15.06
TOTAL	100

It was deemed a representative paper waste composition because Eastleigh has a good mix of socio-economic groups. Moreover, data of KW, GW and PW are given for summer and winter months, allowing us to see the variation (mainly in garden waste) over the seasons and to calculate an annual average. Finally, Eastleigh has a well-established dry recyclables collection of card, newspaper, cans, bottles etc and the data is set out so that actual refuse (Total MSW minus recyclables) is available.

This is desirable as legislation says that all city councils in the UK now have to pick up at least two dry recyclables in their waste collections, meaning that Eastleigh refuse MSW is representative of future MSW in the UK. The TS content of the mixture of waste was adjusted with tap water to obtain the OFMSW feedstock at 10% TS. In other words, to prepare 1kg of feedstock at 10%TS, about 830 g of tap water was blended with 18.6g of GW, 71g of KW and 82g of PW.

2.5.2. Properties

2.5.2.1. Physico-Chemical Analysis

The SCOD for the KW, GW and PW was obtained by drying the waste (105°C) overnight, grinding to obtain a powder and placing it in a dessicator for one day. A suspension was thereafter prepared by mixing 2 g of waste per liter in a blender. The suspension was then mixed with a magnetic stirrer for one day at ambient temperature and then filtered through a paper filter (Fisherbrand QT280).

The low SCOD/COD ratios in Table 5 show the poor tendency to dissolve and this tendency is even worse (8.4%) for the OFMSW. It is hypothesized that most of the soluble material originates from the KW which has the highest SCOD/COD ratio. Nopharatana et al. (2006) found a SCOD/TCOD ratio equal to 7% in a similar experiment. Moreover, Table 5 reveals that the COD/VS ratio for the feed is about 1.3 g O_2/g VS, which is close to values found in the literature for OFMSW: 1.4 (Hartmann and Ahring, 2005; Nopharatana et al., 2006), 1.284 (Anderson and Saw, 1992). The high ash content of garden waste could be explained by the presence of sand and stones among the organic material.

Table 5. Physical and chemical properties of the simulated OFMSW and its components used in this study. (Standard deviation is indicated in brackets for sample in triplicate). dm = dry matter

	Kitchen Wastes	**Garden Wastes**	**Paper Wastes**	**Simulated OFMSW**
Gravimetric tests				
Total Solids (%)	22.9	33.7	94.4	53.8 (1.3)
Volatile Solids (% of TS)	95.4	66.5	85.5	84.4 (1)
Ash (% of TS)	4.6	33.5	14.5	14.7 (1.2)
COD tests				
COD (mg O_2/gdm)	1,369.7 (135)	870.1 (45)	1,330 (61)	1,089.4 (120)
SCOD (mg O_2/gdm)	471.2 (5)	71.7 (4)	15.9 (2)	91.1 (7)
COD (mg O_2/g VS)	1,462.8 (279)	1,041.5 (88)	1,555.7 (72)	1,331.2 (155)
SCOD (mg O_2/g VS)	380.5 (17)	136 (5)	18.5 (2)	111 (10)
SCOD/COD Ratio (%)	34.3 (3)	8.2 (0.6)	1.2 (0.2)	8.4 (1.1)

In addition to COD tests, carbohydrate and protein concentrations were measured on filtered (0.45 μm filter) and unfiltered samples of the simulated OFMSW feedstock. For the unfiltered samples, carbohydrates, on a COD basis, represent 78%, whereas proteins represent only 7%, as shown in Table 6, while the remaining fraction may be aromatic compounds, lipids, humic and fulvic acids, etc. For the filtered samples (≤ 0.45 micron), the proportion of carbohydrates is only 1.7% which implies that carbohydrates from the OFMSW are huge and complex molecules, of which a tiny fraction is composed of small molecules that are solubilised. This is in line with Nopharatana et al. (2006) who found that only 2% of the SCOD was made up of the simple sugars, glucose and maltose, and VFAs like lactic and acetic acids. Table 6 also implies that about 95% of the soluble fraction are not carbohydrates, nor proteins which means that most of the soluble fraction consists of complex molecules which must be broken down into simpler molecules before they can be taken up by microorganisms. Eventually, other relevant properties of the feedstock such as pH, total nitrogen (TN) and total phosphorus (TP) were also measured in order to characterize the simulated feedstock of the OFMSW. The pH was found to be in the range 6.3 - 6.9. TN and TP were found to be 13.1 and 1.3 mg/g TS, respectively, whereas the respective standard deviations were 0.4 and 0.1.

Table 6. Concentration and percentage of carbohydrates and proteins in total and soluble fractions of OFMSW (standard deviation is indicated in brackets)

Unfiltered sample	**Carbohydrates**	**Proteins**	**COD**
mg/g VS	976.9 (59)	62.3 (11)	1,331.2 (155)
% of TCOD	78.5 (8)	7 (1)	100
Filtered sample (≤ 0.45μm)			
mg/g VS	1.8 (0.1)	2.7 (0.2)	111.3 (10)
% of TCOD	1.7 (0.1)	3.7 (0.3)	100

2.5.2.2. Biochemical Methane Potential Test

The test was was carried out with fresh screened (≤ 500 μm) anaerobic sludge from a conventional sewage sludge anaerobic digester (Mogden, UK). The biomass TSS concentration was 34.4 ± 0.6 g/L and VSS was 21 ± 0.5 g/L. The BMP test was determined in duplicates in 38 mL serum bottles, with 20 mL of media at an initial I/S ratio of 2.1, except for the feedstock for which it was 0.7. Some authors (Lay et al., 1997; Nopharatana et al., 2006) used the Gompertz equation that describes the cumulative methane production in batch assays.

$$M = P.exp\left\{-exp\left[\frac{R_m.\exp(1)}{P}.(\lambda - t) + 1\right]\right\} \quad (2)$$

where: M is the cumulative methane production (mL), P is the methane production potential (mL), R_m the maximum methane production rate ($mL.d^{-1}$), λ is the duration of lag phase (d) and t is the duration of the assay at which cumulative methane production M is calculated (d). The parameters P, λ and R_m were estimated by applying a least squares fit of the above equation to the experimental data set using the Solver in Excel.

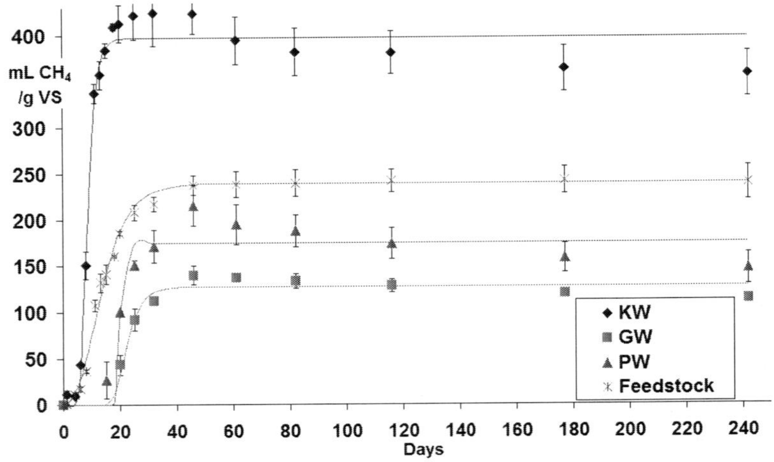

Figure 5. Biochemical Methane Potential on the components and the feedstock of OFMSW used in this study. markers: average of measurements; line: fitted using Gompertz Equation; error bars: standard deviation of triplicates.

Figure 5 shows that all wastes display a lag phase due to the non-acclimatised inoculum used for the experiment. The biostabilization of the organic substrate in anaerobic batch reactors is known to display a slow phase which is the necessary period for adaptation and beginning of bacterial mass multiplication, and a rapid decomposition phase when the bacterial growth is the highest (Lopes et al., 2004). The period of the lag phase is mainly influenced by the initial concentration of microorganisms and substances stimulating the growth of bacteria (Lay et al., 1997). The lag phase, λ, however, is shorter for KW (5.8 days) and Feedstock (4.5 days) than for PW (18.4 days) and GW (17.5 days). This lag phase may be explained by the close association between cellulose and lignin in PW and GW, which impedes access to cellulose. This association might prevent bacteria having direct access to hydrolyzable material which could explain why the blank bottles produced more gas at the very beginning of the test. This translated to negative values for PW and GW during the first twenty days. Interestingly, the lag phase for the Feedstock was less than the one for KW. This is possibly due to the fact that the Feedstock was blended to reduce the particle size and to take a homogeneous sample, and not the

KW, or due to a better C/N ratio for the combined waste. The hydrolysis might have thus been enhanced for the feedstock, and gas production might have started earlier.

Kitchen Wastes are readily hydrolysed and converted to methane. The ultimate methane potential was 357 mL CH_4/g VS, which corresponds to 66% (± 7) of the theoretical methane production based on its COD content. After 50 days, the bottle containing KW stopped producing methane while blank bottles were still producing a little bit which explains why the curve decreased. This biodegradability is consistent with Shin et al. (1993), Lee et al. (1997) and Zhang et al. (2007) who found 356, 343 and 348 mL CH_4/g VS, respectively.

Garden wastes are slowly biodegraded due to their high lignin content which is not degradable under anaerobic conditions. The ultimate methane potential was estimated to be 113 mL CH_4/g VS, which is 33.1% (± 3) of the theoretical methane production based on the COD content of that waste. This value is consistent with the 124 mL CH_4/g VS found by Owens and Chynoweth (1993).

Paper waste is heterogeneous because it is made up of easily degradable paper such as office paper, but also less degradable paper such as newspaper. It is, however, degraded faster than garden wastes and eventually reached 146.6 mL CH_4/g VS, which corresponds to 27.9% (± 2) of the theoretical methane production on a COD basis. The value of ultimate methane production was close to the value reported in the literature (Owens and Chynoweth, 1993).

The feedstock had an intermediate behaviour, as expected because it consisted of a mixture of the three wastes; the ultimate methane potential was estimated to be 239.1 mL CH_4/g VS, which is 45.5% (± 2) of the theoretical methane production based on the COD content of the feedstock. The methane yield was consistent with values reported in the literature. For instance, Nopharatana et al. (2006) found 0.24 L CH_4/g VS. As a conclusion to this test, the BMP revealed that 46% of COD of the feedstock was converted to methane, which is close to the value found by Nopharatana et al. (2006) who carried out a BMP test on the OFMSW (48%). Moreover, 87% of the total methane production was obtained in 32 days. Better performance could probably be obtained with an acclimatised biomass. After 177 days, all the bottles contained a gas in the head space with a methane percentage comprised between 50 and 60%. All the experimental results and the Gompertz coefficients are gathered in Table 7.

Table 7. Gompertz coefficients and experimental ultimate biodegradabilities

Waste	Gompertz coefficients			BMP	biodegradability
	P (mL CH_4)	R_m (ml CH_4/day)	λ (day)	mL CH_4/g VS	%
KW	397.1	72.6	5.8	357.1	66
GW	127.4	13.6	17.5	113.7	33.1
PW	175.2	63.1	18.4	146.6	27.9
Feedstock	237	13.8	4.5	243.8	46

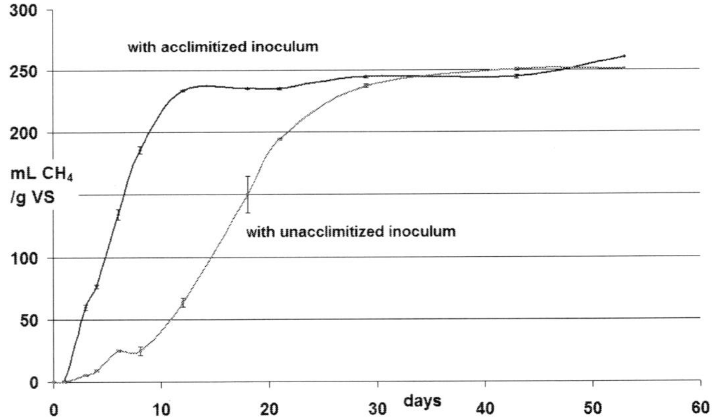

Figure 6. Effect of the acclimatisation of the biomass on the Biochemical Methane Potential test of OFMSW. The error bars show the standard deviation.

2.5.2.3. Effect of Biomass Acclimatisation on the Yield

Another BMP test was carried out on the OFMSW to determine the influence of acclimatisation on the methane production rate. The acclimatised biomass was taken from a 4L chemostat batch-fed with leachate from OFMSW and operating at 20 days HRT; the mixing was stopped 2 days prior to effluent removal in order to let the biomass settle and to keep it within the chemostat. The leachate was produced in another 4L chemostat batch fed with the OFMSW at an OLR of 1 g VS/L.day. The unacclimatised biomass was taken from a drum containing anaerobic sludge from a conventional sewage sludge anaerobic digester (Mogden, UK). In order to compare the two inocula, their VSS content was adjusted to 0.5 g VSS/L and the BMP test was conducted in duplicates at an initial I/S ratio of 0.25.

Figure 6 shows that an acclimatised biomass resulted in a similar methane potential. However, the lag phase (obtained by ~~the~~ Gompertz equation) was found to be significantly shorter for the acclimatised biomass: 1.4 days versus 7 days. It is suggested that the acclimatised biomass includes microorganisms producing more enzymes compared to the unacclimatised ones. It can be seen that the unacclimatised inoculum gives a sluggish methane production rate but eventually both curves reached similar ultimate biodegradability circa 250 ml CH_4/g VS. Based on the COD content, the biodegradability was 49.6 and 47.7% for the acclimatised and the unacclimatised inocula, respectively. The biochemical methane potential assay can be used to determine the hydrolysis constant if the biodegradation kinetics is assumed to follow a first-order reaction according to (Lee et al., 2009):

$$r = -k \cdot S \tag{3}$$

where k is the reaction rate constant (d^{-1}), S is the substrate concentration (g VS/L) and t is the time (days). Integrating this equation with initial and final boundary conditions,

$$ln\left(\frac{S}{S_0}\right) = -k.t \tag{4}$$

The production of methane is related to the substrate degradation by

$$\frac{S}{S_0} = \frac{Y_{max} - Y}{Y_{max}} \tag{5}$$

where Y is the cumulative methane yield at time t and Y_{max} is the ultimate methane yield. From the two equations above:

$$ln\left(\frac{Y_{max} - Y}{Y_{max}}\right) = k.t \tag{6}$$

Using the data from the BMP using acclimatised inoculum and by plotting the equation above, the slope gave the value of the hydrolysis constant as 0.21 d^{-1} ($R^2 = 0.976$). Thus the kinetics of gas production is satisfactorily predicted with the first-order model.

Table 8. Heavy metal concentrations in the simulated MSW (standard deviation is indicated in brackets for sample in triplicate)

Element	ppm of TS
Ca	38,698.66 (4733.6)
Fe	2,325.39 (266.3)
Na	2,256.59 (265.5)
K	1,848.02 (208.5)
Mg	821.34 (93.9)
Zn	588.17 (66.3)
Al	551.94 (66.9)
Mn	88.75 (10)
Cu	47.79 (5.4)
Co	45.16 (5.1)
B	25.73 (4.4)
Pb	8.93 (3.6)
W	6.83 (0.9)
Ni	4.73 (0.9)
Cr	4.2 (0.5)
Cd	2.1 (0.2)
V	< 2.6
Mo	< 2.6

2.5.2.4. Heavy Metals

The heavy metals content of the OFMSW used in this study are listed in Table 8. The high content of calcium, iron, sodium and potassium can be pointed out in this table. These are presumably due to KW. Zhang et al. (2007) carried out an elemental analysis of food waste and found concentrations of 21,600 ppm calcium, 766 ppm iron, 900 ppm potassium and 1,400 ppm magnesium.

3. Results and Discussion

3.1. Performance of the Hydrolytic Reactor

3.1.1. Effect of the OLR on VS Removal

The TCOD in the leachate varied over a wide range, between 4000 and 26,000 mg/l, possibly due to the HR being fed every two days until day 159, intermittent mixing, and due to occasional stirring difficulties. It can be seen from Figure 7 that the TCOD did not change with changes in OLRs from 0.5 to 16 g VS/L.day.

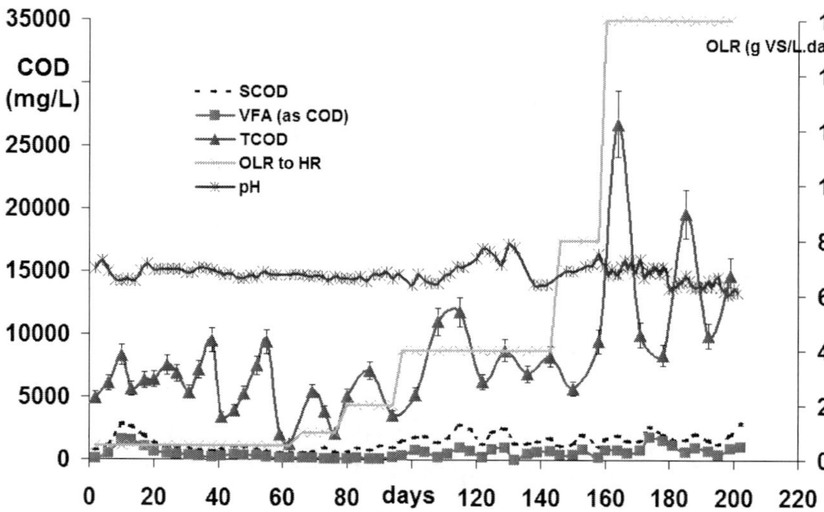

Figure 7. Evolution with time of TCOD, SCOD and VFAs in the effluent of the HR (left axis), and OLR and pH (right axis). Reprinted from (Trzcinski and Stuckey, 2009b).

However, the value of TCOD did depend upon the occasional presence of solid particles in the sampling line at the time of sampling. Similarly, the SCOD did not vary significantly when a step increase in OLR was effected in the HR, and was always in the range 530 – 2,900 mg/L. The low SCOD observed in the leachate is thought to be due to poor hydrolysis because of the ligno-cellulosic type of feedstock, and an inadequate amount of inoculum used to seed the HR.

The initial inoculum to substrate ratio (I/S) may also play a role during start-up of anaerobic digesters. The initial I/S ratio was 0.02 based on the initial load of 340 g volatile solids fed during start up. Then the HR was fed continuously at an OLR of 0.5 g VS/L.day but with intermittent mixing as well as occasional stirring difficulties at TS above 5%. Table 9 presents the VS removal percentages at the various OLRs and HRTs tested. The VS removal percentage was calculated as follows:

$$\text{VS removal\%} = 100\% \left(1 - \frac{mass\ VS\ removed + mass\ VS\ accumulated\ in\ HR}{mass\ VS\ fed\ in\ HR}\right) \tag{7}$$

Where the masses were considered over a period longer than 15 days so that steady-state can be assumed and the mass of VS accumulated in the HR is the difference between the mass of VS in the HR at the beginning and the end of the period considered. The VS removal percentages shown in Table 9 are 65.4, 43.8, 35.5, 22 and 13.8% VS destruction at 0.5, 2, 4, 8 and 16 g VS/L.day, respectively, assuming that the volatile solids production due to bacterial growth and the transfer of volatile solids to the SAMBR were negligible.

The transfer of volatile solids to the SAMBR was very limited thanks to the separation between coarse solids and leachate by the perforated stainless steel mesh within the HR. Nevertheless, a small fraction of solids could still pass through and be pumped to the SAMBRs. At the end of the experiment large holes and rust were observed in the 50 micron stainless steel presumably due to a low quality steel or due to pieces of wood in the GW that may have pierced the mesh. This transferred VS fraction over 200 days was estimated as 37.8 and 69.3 g VS for SAMBR1 and SAMBR2, respectively, which can be considered as negligible. For instance, during the period at 16g VS/L.day (day 159 to day 199) the total VS mass transferred to SAMBR1 and SAMBR2 together equalled 91 g changing the VS removal % in the HR to 12.4 instead of 13.8. The former is the actual VS removal in the HR, while the later could be named the "apparent VS removal" and in this study they were similar and thus the difference was neglected. The low VS removal percentages were also due the low volatile solids retention times calculated as the ratio of mass of volatile solids in the HR that is equal to X · V, where X is the VS concentration in g/L in the HR and V is the HR working volume in L, and the mass of volatile solids removed from the HR per day (W in g VS/day):

$$\text{VS Retention Time (days)} = \frac{X \cdot V}{W} \quad (8)$$

Consequently, the anaerobic biodegradability of the digestate that was taken out of the HR was consistent with the lower VS removal observed as the OLR was increased. The BMP of the digestate was carried out at an I/S of 0.1-0.2 using the acclimatised sludge from SAMBR2 as inoculum, and the ultimate biodegradability was found to be 167.7, 229.7 and 296.6 mL CH_4/g VS fed at OLRs of 0.5, 8 and 16 g VS/L.day, respectively.

Table 9 also contains the HRT of the HR, i.e., the hydraulic retention time or leachate retention time, which is the average retention time of a unit volume of liquid in the reactor and is calculated as the ratio of the reactor volume and the leachate flowrate to the SAMBRs. Obviously, the lower the HRT, the higher the flowrate of leachate to the SAMBRs and the more likely the acidification of the HR can take place because methanogens do not have time to grow in sufficient amount to degrade the VFAs in the HR. However, it is possible that the mesh inside the HR acts as a bacterial support where methanogens can attach and form colonies with acidogens so that they can survive on VFAs produced in the close environment. Figure 8 shows the methane content of the biogas produced in each reactor. It can be seen that methane started to be produced very slowly compared to the SAMBR.

Table 9. Comparison of volatile solids retention times, volatile solids removal percentages, fresh water consumption, hydraulic retention times, percentages of fresh water addition compared to recycled process water and digestate, methane potential at different organic loading rates in the hydrolytic reactor. n.a. = not applicable

OLR (g VS/L.day)	0.5	2	4	8	16
Duration (days)	63	17	47	14	40
VS RT (days)	67.8	49	16.6	6.4	3.3
VS removal (%)	65.4	43.8	35.5	22	13.8
Average fresh water consumption (mL/day)	3.7	n.a.	68	202	652
HRT (days)	15	9	7.8	4	2.2
% of fresh water added compared to recycled process water	0.6	n.a.	5	8	14
Digestate methane Potential (mL CH_4/g VS)	167±6.2	n.a.	n.a.	229.7±6.9	296.6±24

The methane percentage increased gradually to circa 40% around day 100, but on day 112 the HR was opened and the 50 microns macrofilter was removed and replaced by a new one which let some air into in the headspace. As a result, the methane content dropped and remained low as the reactor had to be opened more frequently due to technical issues such as mixing problems or solids that could not be removed from the bottom pipe due to the high TS content.

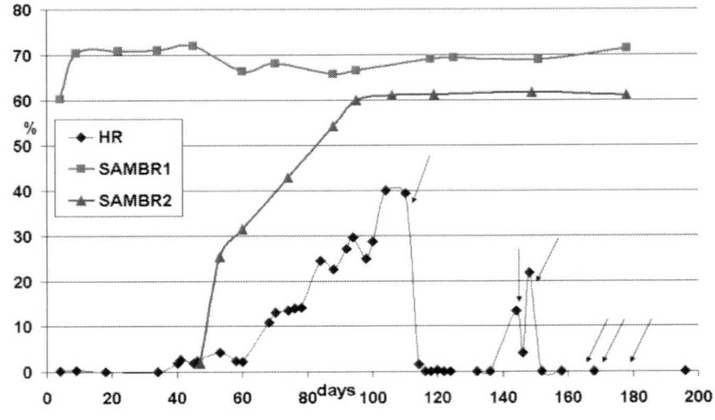

Figure 8. Evolution of the methane content in the HR, SAMBR1 and SAMBR2. The arrows show when the HR was opened for maintenance work on the macrofilter.

3.1.2. Effect of the OLR on VFAs Production and Composition

The evolution and composition of VFAs over time in Figure 9 shows that acetate was the main VFA at steady-state, but propionate became the main acid after the shock at 4, 8 and 16 g VS/L.day on days 101, 146 and 164, respectively, which is a few days after the organic shocks took place (Trzcinski and Stuckey, 2009b). The HRT also affects the VFAs distribution because it was reduced to 4 and 2 days on day 146 and 164, respectively. From day 160 onwards, the HR was fed every day at 16 g VS/L.day at HRTs ranging from 4 to 1 days, and propionate remained the main acid until the end of the run.

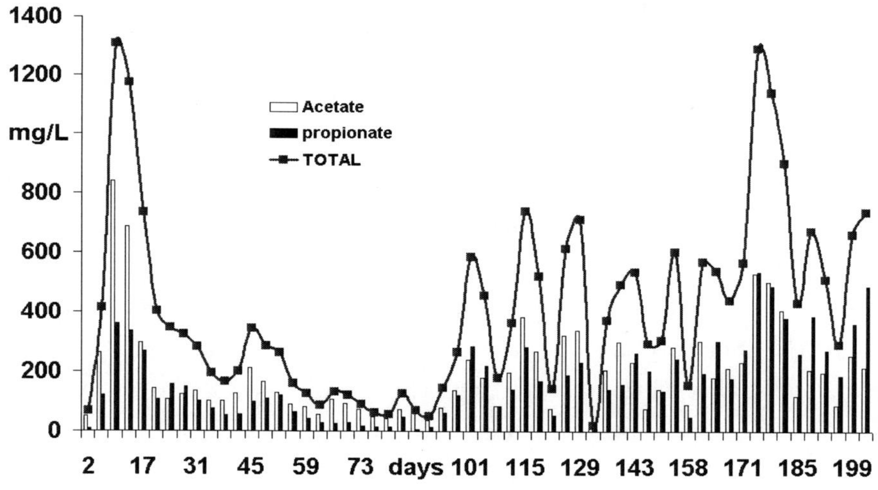

Figure 9. VFA distribution in the effluent of the HR. Reprinted from (Trzcinski and Stuckey, 2009b).

Gallert et al. (2003) also observed a higher and longer-lasting propionate accumulation when the HRT was reduced from 7.1 days to 5.7 days at an OLR of 15 kg COD/m^3.day; they correlated this with 1% hydrogen in the off-gases. Propionate oxidation is known to be the bottle neck reaction during the methanogenesis of complex substrates because their growth rate is only 0.13 d^{-1} (Wallrabenstein et al., 1995), and they can be washed out at a HRT below 8 days (Gallert et al., 2003). The pH dropped between 6 and 6.5 due to the OLR of 16 gVS/L.day applied after day 159, but with the accumulated alkalinity (5,000 mg equivalent $CaCO_3$/L on day 199) and the recycling of the SAMBR1 and AMBR permeates, the pH did not drop any further, which highlights the advantage of recycling the SAMBR permeate. A BMP test (I/S = 1.3) on the leachate on day 193 revealed that 57% (± 10%) of the leachate was converted to methane. The rest of the COD was either not hydrolysable or converted to biomass.

3.1.3. Scanning Electron Microscopy of the Fibers in the HR

In order to gain more understanding of the hydrolysis process, a Scanning Electron Microscope was used to visualize the fibres of the OFMSW. Since the KW is easily degraded, and since GW represented a small percentage of the feedstock, it can be assumed that the fibres originated from the PW. Different magnifications were taken of both the virgin fibres taken from the feedstock, and the fibres from the HR after anaerobic digestion. All the pictures are shown in Figure 10. The virgin fibres are on the left-hand side (Pictures A, C, E and G), and the treated fibres are on the right-hand side (Pictures B, D, F and H). The pictures show evidence that the fibres are being peeled off and small nodules looking like bacteria could be seen on the surface (Figure 11). The surface is much rougher than virgin fibres which is due to bacterial hydrolysis that targets the cellulose and hemicellulose from the fibrils, but it could also be due to the shear of the stirrer in the HR. Figures B and D show that pieces up to 10 microns can be peeled from the main fibre, but still remain attached to it. This could be the result of a partial hydrolysis of cellulose that was consumed until only lignin remained accessible to enzymes. Some holes were observed in the virgin fibres of picture E. This could be specific to a certain type of fibre (cardboard, newspaper, office paper, etc.) or it could be the result of solubilisation of hemicellulose during the autoclaving of the virgin fibres. Solubilisation of hemicellulose has indeed been reported possible in hot water (Hendriks and Zeeman, 2009). Liao et al. (2004) also observed holes in manure fibres before acid treatment and they attributed this to a partial degradation of cellulose.

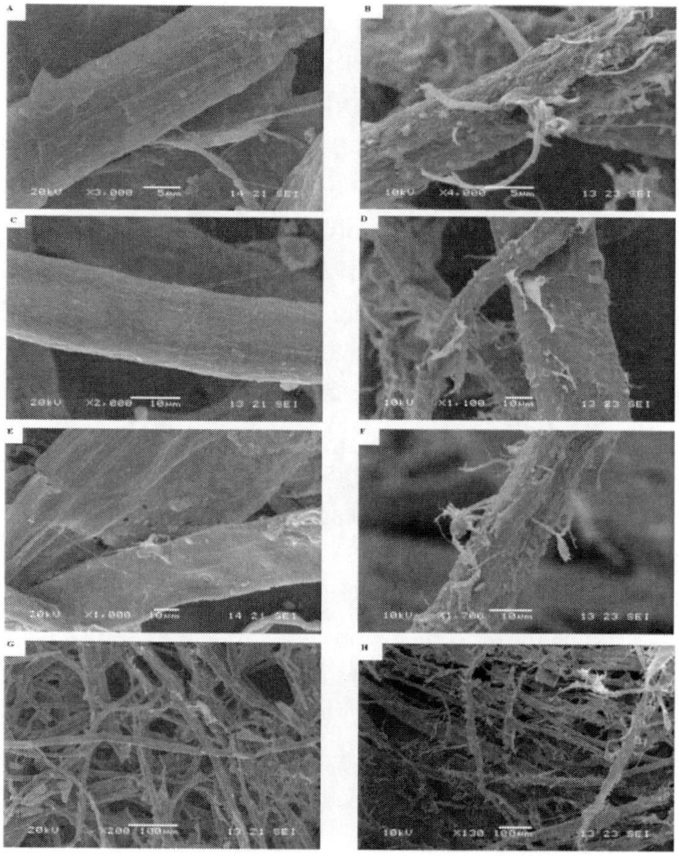

Figure 10. Scanning Electron Microscopy of the raw fibres from the OFMSW feedstock (left) and treated fibres from the HR (right).

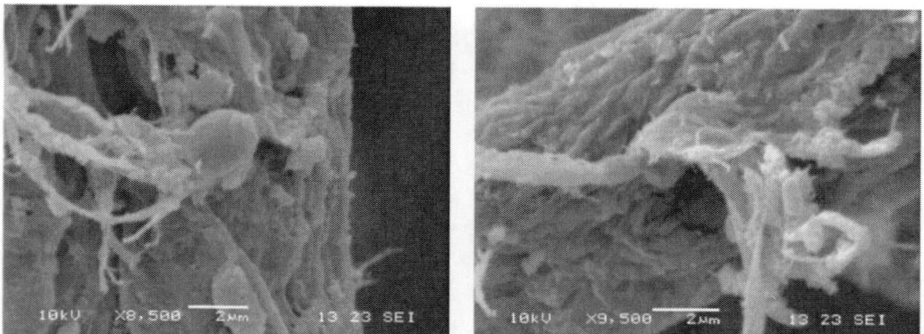

Figure 11. Close zoom on the fibres treated in the HR. This is probably bacterial aggregates attached to the fibre.

3.2. Performance of SAMBR1

3.2.1. Effect of the HRT on COD Removal

The OLR to the SAMBR was not constant because of the fluctuations in the TCOD of the leachate from the HR (Figure 7), and as a result the SCOD in SAMBR1 (Figure 12) sometimes increased sharply over time. For instance, an OLR to the SAMBR of 8 g COD/L.day was observed temporarily on day 164, and a decrease of the HRT to 2.1 days led to a sharp peak of SCOD in the reactor but not due to VFAs build-up, indicating that the hydrolysis was rate limiting. On day 185, a temporary OLR of 19.8 g COD/L.day was also observed, but the stability of SAMBR1 at such a high OLR could not be assessed properly due to the transient character of the OLR. Despite the varying OLR, the permeate SCOD (effluent SCOD in Figure 12) remained typically in the range 300-500 mg /L. A build-up of the SCOD was not observed and the SCOD was even found to decrease slightly on few occasions. This can be partly attributed to the greater consumption of fresh water towards the end of the run to keep up the volume in the HR (see Table 9), but the decline of SCOD was also due to the very high MLTSS (28.7 g/L) at the end of the run, and was not due to the enhanced rejection by the membrane because the SCOD in the bulk liquid was also found to decrease slowly.

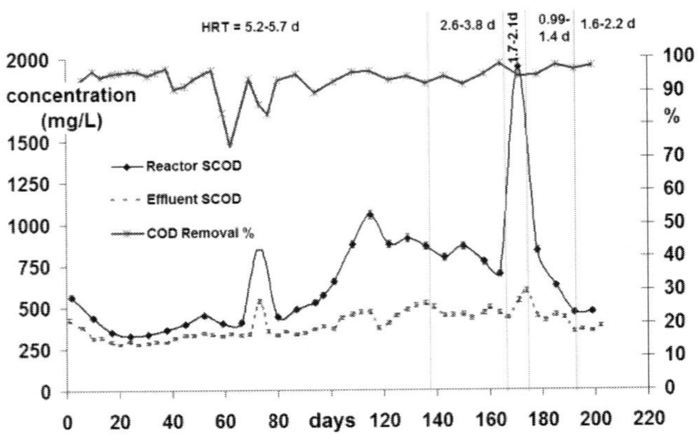

Figure 12. SCOD and VFAs inside SAMBR1 and in its permeate (left axis). COD removal in SAMBR1 (right axis). Reprinted from (Trzcinski and Stuckey, 2009b).

3.2.2. Effect of the TSS on the Flux

The COD removal remained circa 95% throughout the run, while the VFA concentration was virtually zero, suggesting that the methanogenic population could cope with an HRT as low as 1 day. However, SAMBR1 could not be operated in a sustainable way at a HRT below 1.6-2.3 days due to a membrane flux limitation at 0.54-0.78 LMH. At an HRT below 2 days, the rate of particulate COD destruction became less than the feeding rate, resulting in the build-up of solids at the bottom of the reactor which eventually blocked the diffuser and, on day 182 no more bubbles were scouring the membrane. At the same time, the MLTSS increased to 28.7 g/L (Figure 13) which also badly affected the flux. This indicates that the performance of the SAMBR treating leachate containing particles was limited to 1.6-2.3 days HRT by the hydrolysis and not the VFA degradation. Interestingly, as the HRT dropped the inorganics in suspension followed the same trend as TSS and VSS with a sharp increase to 13.6 g/L showing that at HRTs lower than 1 day the FSS build up in SAMBR. Again because the hydrolysis of solids is too slow compared to the feeding rate, the inorganics bound to the ligno-cellulosic matrix are not solubilised fast enough and build up. As a result the VSS/TSS ratio plummeted to circa 50%. The transmembrane pressure (TMP) in SAMBR1 was virtually zero until day 60.

SAMBR1 was treating leachate from day 0 to day 60 at a HRT of 5.5 day, at a stable flux of 8.5 LMH and a MLTSS around 5 g/l with virtually zero TMP. The MLTSS could remain that low because the HRT was 5.5 days which provide enough time for the hydrolysis to proceed without noticeable build up in MLTSS. On day 24, it was attempted to increase the flux by increasing the speed of the recycling pump but that led to a sharp increase in TMP to 450 mbar which indicates that operation over the critical flux was taking place and that a cake layer was forming. Based on the step increase in the recycling pump setting, it could be approximated that the flux was near 11 LMH which was thus deemed over the critical flux. Between day 76 and day 105, the effect of the sparging rate on the COD rejection was investigated using TMP data which were analysed in the corresponding section. After day 105, the flux in SAMBR1 remained relatively stable at 3.5 LMH although the MLTSS varied over a wide range (4 - 19 g/L). This can be explained by the increase in sparging rate to 10 LPM that results in the resuspension of solids that settled down in the reactor at low sparging rate.

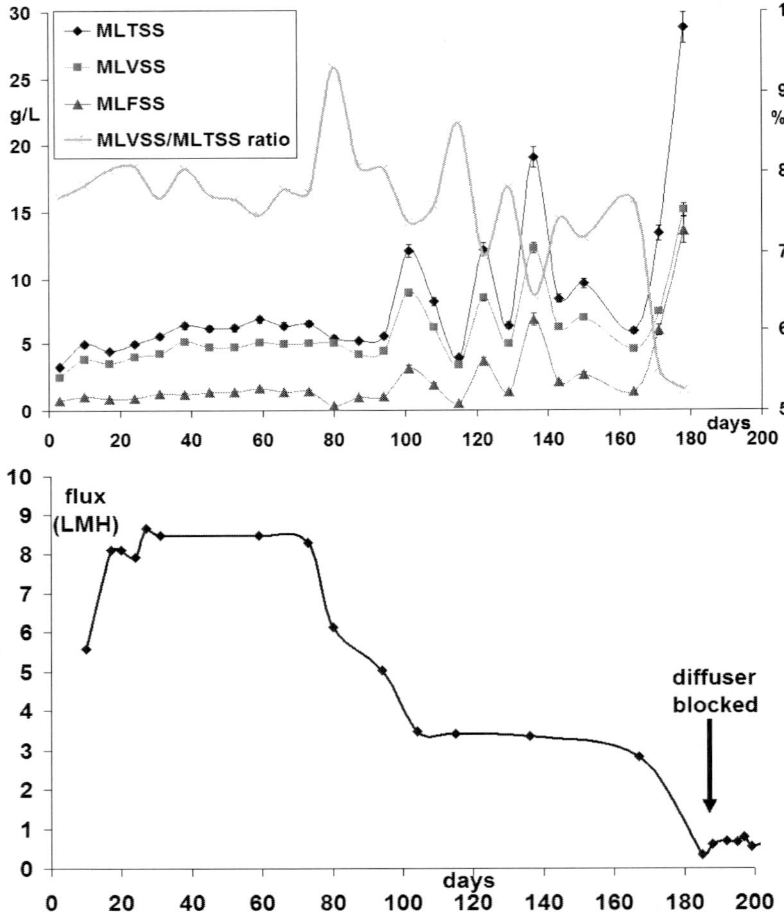

Figure 13. Evolution with time of the TSS, VSS, FSS and VSS/TSS ratio (top), and evolution of the flux in SAMBR1 (bottom).

On a few occasions, however, the sparging rate had to be reduced due to foaming problems which caused the solids in suspension to settle down and this resulted in a lower apparent MLTSS. With this in mind, the great variation in MLTSS should not be seen as a result of changing performance in the reactor but rather as non-homogeneous sampling for MLTSS measurement. The flux of 3.5 LMH was accompanied by TMP values of up to 500 mbar whereas the operation after day 170 at a MLTSS of 28.7 was characterized by very low flux circa 0.5 LMH and very high TMPs in the range 550 - 800 mbar (data not shown). Normal operation of the Kubota membrane is associated with fluxes in the range 20-27 LMH at a maximum of 15 mbar TMP and 20 g TSS/L (Ramphao et al., 2004), but Van Zyl et al. (2007) reported a flux of 2-3 LMH at 36 g TSS/L when treating synthetic petrochemical effluent which is similar to what was observed in this study although their TMP was much lower (50 mbar) probably due to the feed containing only soluble organics. At such high MLTSS, it is thought that the diffuser was blocked due to the solids settling at the bottom of the reactor. This led to a much reduced scouring of the membrane. In order to lower the MLTSS and improve the flux, sludge was removed from SAMBR1 at a mean rate of 90 mL/day after day 180 (SRT=33 days), but this did not help in reducing the MLTSS.

3.2.3. Effect of the Bulk COD on Membrane Rejection

The SCOD inside the reactor remained higher than the effluent values throughout the experiment, which demonstrates that the presence of a cake/gel layer on the membrane surface improves the effluent quality: this is in line with previous work on the SAMBR (Akram, 2006; Choi and Ng, 2008). Nevertheless, membrane rejection did not increase with time but varied according to the bulk SCOD. The rejection of solute can be due to the dynamic layer formed by the foulant, and the attached biomass acts as a secondary filter. Choi and Ng (2008) observed that solute rejection rate increased with more severe fouling. During fouling, pore blocking and pore narrowing can take place which would enhance solute rejection. They also observed lower TOC in the permeate compared to the supernatant, which they attributed to a combination of biodegradation by the biofilm (cake layer) developed on the membrane surface and further filtration by cake layer and narrowed pores. Hence, lower permeate concentration and faster permeability decline are observed as membranes are fouled. Membrane rejection was expressed as a percentage:

$$\text{Rejection} = 100\% \cdot \frac{SCOD_{bulk} - SCOD_{permeate}}{SCOD_{bulk}} \quad (9)$$

In this study it was observed that the higher the bulk SCOD, the higher the rejection (Figure 14), which suggests that the high molecular weight COD is degraded in the bulk liquid until it becomes small enough to pass through the membrane or the biofilm pores.

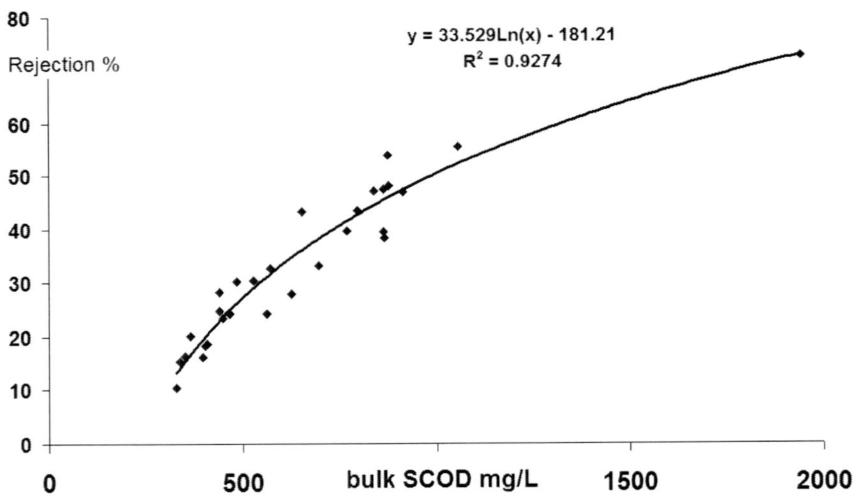

Figure 14. Correlation between the bulk SCOD in SAMBR1 and the membrane rejection. Reprinted from (Trzcinski and Stuckey, 2009b).

3.3. Performance of SAMBR2

3.3.1. Effect of Inoculum on Start-Up

Previous studies (Akram, 2006) have shown that a shorter start-up period and higher COD removal in SAMBRs can be obtained by increasing organic load at a lower constant HRT rather than gradually decreasing the HRT at constant high feed strength. This approach was followed to start up a SAMBR, although Akram (2006) used a sucrose-based wastewater that is easily degradable, while the leachate used in this study was lignocellulose-based. For an easily degradable substrate, VFA accumulation can occur in the SAMBR due to the overloading of the methanogens and possibly the lack of syntrophic associations necessary to degrade reduced intermediates. For this reason, prior inoculation into a CSTR is helpful for the development of an active inoculum enriched in methanogens (Akram, 2006). With this in mind, an inoculum was fed on synthetic VFAs as the main carbon source (75% on a COD basis) in a 4 litre chemostat prior to inoculating SAMBR2. Prior to inoculating SAMBR2, a specific acidogenic activity test was conducted on the two different inocula, the one from SAMBR1 and the one from the chemostat batch fed with meat extract, peptone and synthetic VFAs. The test was determined in triplicate in 38 mL serum bottles, with 20 mL of media. The same amount of glucose was fed to both sets of bottles to result in 2 g COD/L for the test at an I/S ratio of 1, and Figure 15 reveals that indeed the acidogenic and methanogenic biomass of the inoculum fed with synthetic VFAs was more active than the inoculum taken from SAMBR1 on a same MLVSS basis. Akram (2006) also reported a higher activity from an inoculum taken from a CSTR compared to the SAMBR. The reason for this is the bigger flocks formed in a gently mixed CSTR compared to the SAMBR where the shear rate is more important. Several authors have identified strong mixing conditions and high shear forces as a detrimental factor for VFA degradation since it supposedly affects microbial proximity, and therefore interspecies hydrogen and/or formate transfer (Brockmann and Seyfried, 1997; Kim et al., 2002).

As a result the smaller flocks in the SAMBR can release more hydrogen due to inefficient interspecies hydrogen transfer, and this results in reduced methane production through the hydrogenotrophic reaction. This can also be due to the non-living fraction of MLVSS in the inoculum from SAMBR1 as the MLVSS measurement does not distinguish the non-living VSS from the living fraction. As a result, the inoculum from SAMBR1 was likely to contain less acidogens and methanogens explaining the slower methane production rate during the first two days. It is interesting to note that eventually the inoculum from SAMBR1 ended up producing significantly more methane that the inoculum from the chemostat. With the inoculum from SAMBR1, 79% (±0.6) of the glucose was converted to methane, while that percentage reached 72.3% (± 0.9) with the inoculum from the chemostat. This again can be explained by the amount of VSS introduced into the BMP bottles. As the inoculum from SAMBR1 contained non-living VSS consisting of lignocellulosic fibres, the actual living VSS was less than in the inoculum from the chemostat. Therefore less COD was used for biomass growth and the spare COD could then be metabolized to produce more methane. The extra methane produced could also come from the lignocellulosic fibres that were present in the inoculum from SAMBR1, and that were further hydrolyzed in the serum bottles. As the acidogenic test showed that the inoculum from a chemostat fed on meat extract, peptone and synthetic VFAs was more active on a MLVSS basis, it was hypothetized that it would be beneficial for the start-up of the SAMBR in order to avoid VFA accumulation. Also the smaller inoculum required to seed the SAMBR due to a higher methanogenic activity would result in lower MLTSS concentrations for a successful start-up and maybe a better flux. VFA concentrations in SAMBRs1 and 2 were both virtually zero. This indicates that an inoculum acclimatised to VFAs does not bring further advantages regarding VFAs degradation because both SAMBRs could start-up at a HRT of 5.2-5.7 days with no VFA accumulation. Previous work on leachate from MSW has also shown that microorganisms grown in another medium are unable to out-compete native solid waste microorganisms for the cellulose in a foreign (leachate based) medium (O'Sullivan and Burrell, 2007). In this study, bacteria fed on peptone and meat extract may have been inhibited when suddenly fed with leachate, or merely may have not been able to metabolize lignocellulosic compounds causing a delay in VFAs production; this would explain why the methane content of SAMBR2 displayed such a long lag phase before reaching a normal value of 60% methane in the biogas (Figure 8). Moreover, the methane content of the biogas in SAMBR2 gradually increased to a maximum of 61% after 50 days, whereas in SAMBR1 it reached 60% after four days of operation and then slowly stabilized at values between 69 and 71% (Figure 8), which suggests that the inoculum fed on synthetic VFA was not suitable for start-up because it did not contain hydrolytic and acidogenic bacteria for a leachate medium. As a result, the slow production of VFAs was translated into a slow increase in methane production.

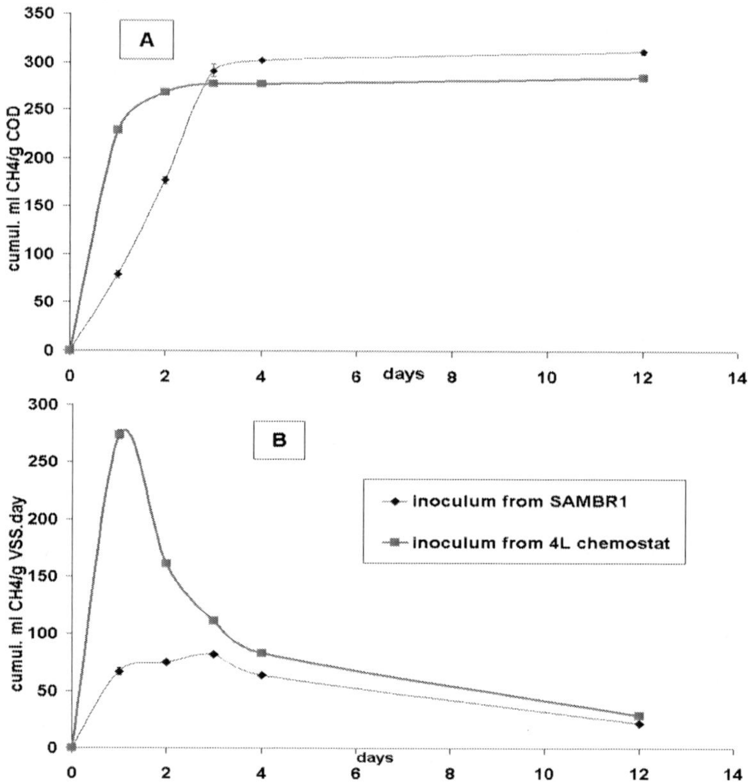

Figure 15. Specific acidogenic activity test on the inoculum from SAMBR1 acclimatised to the leachate medium and the inoculum from a 4 litre chemostat enriched with methanogens in a synthetic medium of peptone, meat extract and VFAs. Panel A: cumulative methane production per g COD fed. Panel B: cumulative methane production per g MLVSS per day. Reprinted from (Trzcinski and Stuckey, 2009b).

However, it could be argued that methane percentage in the headspace is not a good indication of the methane production because it depends on the carbon dioxide solubility, which in turn depends on the pH, alkalinity and buffer capacity. In this case, the pH and the alkalinity (Figure 23) in SAMBR1 and SAMBR2 were very similar which provides a sound prerequisite to compare headspace methane content. Furthermore, the bulk SCOD in SAMBR2 kept increasing during start-up from 100 mg/L on day 46 to 800 mg/L on day 73 after which it began to decrease. This strongly suggests that, for a lignocellulosic-based feed, it is paramount to start up the SAMBR with a hydrolytic population acclimatised to the leachate medium to avoid SCOD build up in the bulk. Eventually, no biogas was produced in SAMBR2 until day 106, whereas the specific methane production in SAMBR1 was in the range 0.11 - 0.18 L CH_4/g COD fed during the first 30 days.

3.3.2. Effect of the HRT on COD Removal

The HRT of the AMBR was equal to the HRT of SAMBR2 because the two reactors were connected in series. The COD removal in SAMBR2 was 94.5% on average, and only 1.6% in the AMBR so that a total COD removal of 96.1% was achieved. The VFA concentration was virtually zero inside SAMBR2 and the permeate, and thus were omitted from Figure 16. No significant change in total COD removal efficiency of both reactors was observed when the HRT was decreased from 5.2-5.7 to 0.37 days. In a moving-bed biofilm reactor system with an anaerobic-aerobic arrangement, Chen et al. (2008a) observed that at 1.5 days HRT the COD removal of the anaerobic reactor dropped to 81%, whereas the aerobic COD removal increased to 11%, but nonetheless the total COD of the system remained stable. Although the contribution of the aerobic step to the total COD removal of the system was low in this study because of the membrane rejection in SAMBR2 (1.6% on average), it should be emphasized that on average 26% of the recalcitrants from SAMBR2 could be degraded aerobically in the AMBR. Bohdziewicz et al. (2001) reported less than 10% aerobic biodegradation of anaerobically treated landfill leachate. The COD in the permeate of the AMBR was approximately 300 mg/L at the end of the experiment, which is close to the 390 mg/L reported by Agdag and Sponza (2005). The SCOD in the bulk of the reactors as well as in their permeate decreased slightly towards the end of the operating period.

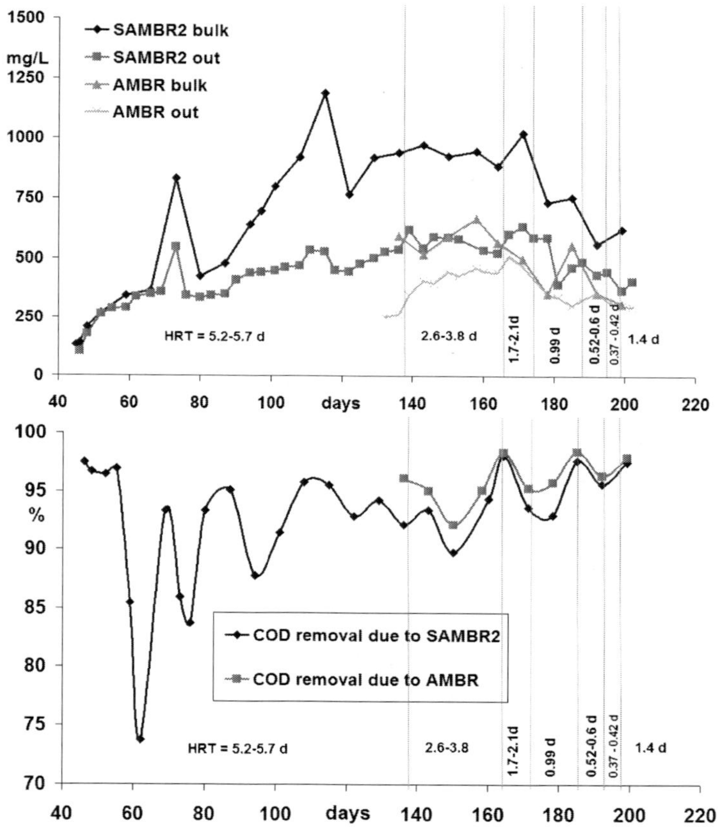

Figure 16. SCOD inside and in the permeate of SAMBR2 (out) and AMBR at different HRTs (top). Reprinted from (Trzcinski and Stuckey, 2009b). COD removal in SAMBR2 and AMBR (bottom).

3.3.3. *Effect of the TSS on Flux*

Figure 17 shows the evolution with time of the TSS, VSS and the FSS. Interestingly, the initial flux in SAMBR2 was much higher than in SAMBR1; the initial flux in SAMBR1 and 2 was 8.6 and 21.6 LMH, respectively although both reactors were started up at similar TSS of 3.3 and 2.6 g/L for SAMBR1 and SAMBR2, respectively. The only difference was in the different inoculum used to seed the SAMBRs. SAMBR2 was inoculated with an inoculum almost free of colloids, whereas SAMBR1 was inoculated with bacteria acclimatised to the leachate medium but containing colloids that are known to be detrimental for the flux. Nonetheless, as SAMBR2 was fed on leachate containing colloids its flux also started to decrease over time. As a result, the TMP started to rise so rapidly that the flux had to be reduced manually. This was the case on day 50 where the TMP rose to 200 mbar, on day 53 where it reached 250 mbar and the flux was reduced to 15.3 LMH. Again, on day 59 the TMP rose to 350 mbar and the flux was reduced to 12 LMH, which stabilized the TMP at 200 mbar. With time the TMP rose again to 300 mbar on day 70 and the flux was reduced to 10.1 LMH to stabilize the TMP between 150 and 200 mbar. This approach was also followed by other researchers (Howell et al., 2004) in order to avoid serious fouling of the membrane.

For relatively ideal wastewater such as synthetic or municipal wastewater, the permeate rate is usually reduced if the TMP exceeds 10kPa (= 100 mbar) because this is equivalent to a 1m liquid head corresponding to the typical liquid depth above the membranes in full size systems. In only 30 days, its flux decreased to the same flux as in SAMBR1 meaning that the advantage of a "colloid free" inoculum was effective only for a short period. As in SAMBR1, the low HRT resulted in the build-up of MLTSS, MLVSS and FSS as well as the drop in the VSS/TSS ratio over time. The operation after day 100 was characterized by an increase in TMP to reach 650 mbar on day 120. The TMP remained constant as well as the flux (around 5 LMH), although the MLTSS increased gradually from 1 to 19 g/L on day 178, which emphasized the independence of the MLTSS on the flux in a certain range of values as reported by other researchers (Hong et al., 2002; Ross et al., 1990; Yamamoto et al., 1989). Then after day 180, the MLTSS went over 20 g/L, and even reached 46 g/L on day 195, which eventually blocked the diffuser and no more bubbles could be seen scouring the membrane, which was correlated with the TMP culminating at 850 mbar and the flux dropping to 0.5 LMH until the end of the experiment. Thus, as in SAMBR1, at HRTs lower than 2 days, particulate solids in the leachate built up at the bottom of the SAMBR eventually leading to the diffuser blocking (see Figure 18). During the last 20 days of the run, sludge from the bulk was removed in an attempt to decrease the MLTSS in SAMBR2. On average, 130 ml were removed per day (SRT =

23 days), but that was not enough to decrease the MLTSS as the feeding rate of particles was too high compared to its removal rate.

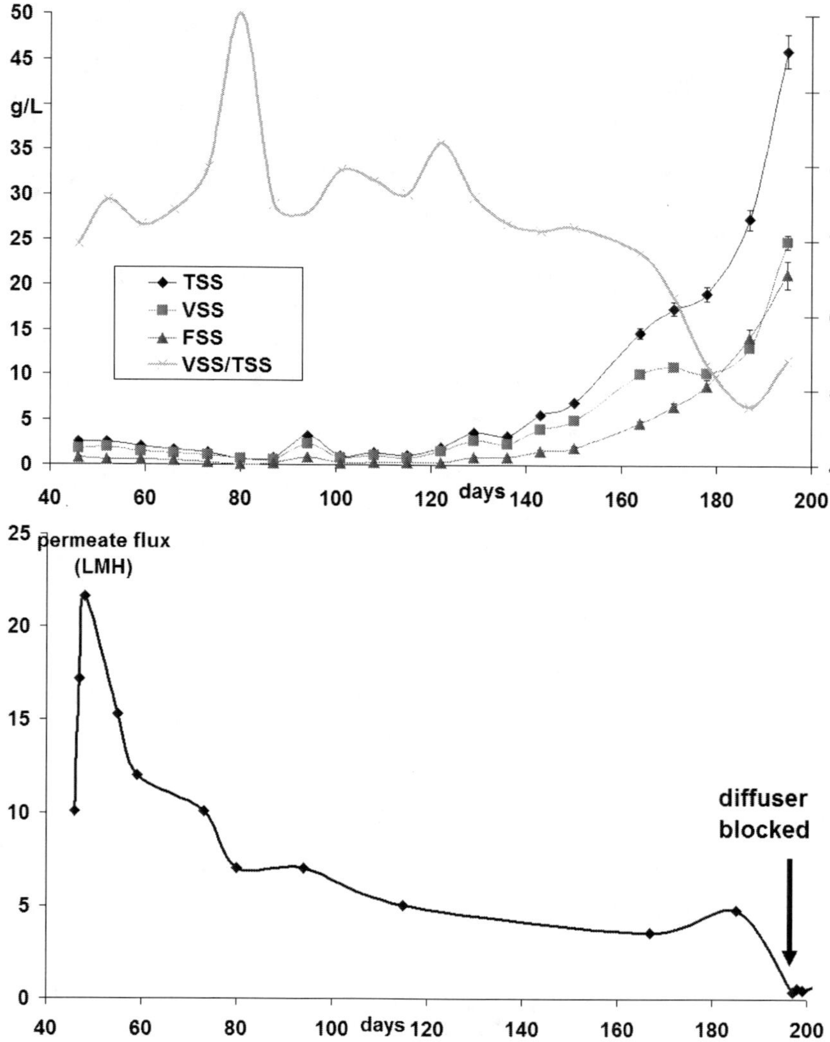

Figure 17. Evolution with time of the TSS, VSS, FSS and VSS/TSS ratio (top), and evolution of the flux in SAMBR2 (bottom).

Figure 18. Left: Kubota membrane module partially scoured by biogas bubbles. Right: blocked diffuser located at the bottom of the SAMBR with only the far left hole remaining open for gas bubbles formation. Reproduced from (Trzcinski and Stuckey, 2009a).

3.3.4. Effect of the Biogas Sparging Rate on COD Rejection

Previous work on the SAMBR has shown that the presence of a biofilm is responsible for a cleaner effluent (Akram, 2006; Choo and Lee, 1996; Harada et al., 1995). Both SAMBRs were fed on leachate at the same HRT of 5.2-5.7 days while the sparging was set at a low value in order to investigate the effect of sparging rate on the rejection of COD by the membrane. Another objective was to see how much time it takes to develop a biofilm and whether this biofilm could result in an enhanced rejection and a lower permeate COD. This could have a significant impact because it would mean that the permeate COD could be controlled by changing the sparging rate and thereby obtain cleaner effluent. On day 76, the two SAMBRs were deemed to be at steady state as they recovered from the sudden shock in OLR that made their COD rise to about 500 mg/L. The sparging rate was then set at 2LPM in both SAMBR. The MLTSS were relatively low and constant around 5 and 1 g/L in SAMBR1 and 2, respectively. It was observed that no significant change occurred and that the permeate SCOD remained stable at around 350-360 mg/L in both reactors. In contrast, the TMP skyrocketed to circa 580 and 380 mbar, in

SAMBR1 and SAMBR2, respectively (Figure 19). This indicated that the flux became greater than the critical flux as soon as the sparging rate was reduced and that a cake layer can form very quickly above the critical flux. The flux was manually reduced to 6.1 and 7 LMH in SAMBR1 and SAMBR2, respectively, to avoid operation at high TMP. The TMP went back to 150 and 100 mbar in SAMBR1 and SAMBR2, respectively, and this TMP was probably not high enough to build a biofilm that would have led to enhanced rejection, and as a result the permeate SCOD did not decrease. Although the flux had been manually reduced to 6.1 LMH in SAMBR1, the TMP slowly increased to 650 mbar at which point the flux was only 5 LMH (measured on day 94). Thus a cake layer had been formed at 6.1 LMH but at a much lower rate than at 8.3 LMH. The rate of increase in TMP indicates the degree of compaction of the biofilm which can translate to denser cake layer and COD rejection. The higher the rate of increase in TMP, the more compact the biofilm will be, and the thicker the biofilm will be. Also the degree of compaction will determine the easiness and the rate at which the biofilm can be removed if the sparging rate is increased.

At very high TMPs, Elmaleh and Abdelmoumni (Elmaleh and Abdelmoumni, 1997) also reported a decrease in permeate flux with an increase in TMP which was attributed to the compaction of the foulant layer. Li et al. (Li et al., 2003) explained that cakes formed at higher flux and TMP are much more consolidated than cakes formed marginally above the critical flux. In SAMBR2, although the flux had been manually reduced to 7 LMH, the TMP rose slowly to 400 mbar before day 87. On day 87, SAMBR 2 was set at a high sparging rate (8-10 LPM), while the sparging rate in SAMBR1 remained unmodified. As a result, the TMP in SAMBR2 dropped to virtually zero, while the permeate SCOD in SAMBR2 increased to about 440 mg/L on day 90 and remained stable while the permeate SCOD in SAMBR1 remained below 400 mg/L. It is interesting to note that the increase of the sparging rate did not allow us to re-establish the initial flux in SAMBR2, although the TMP went back to zero.

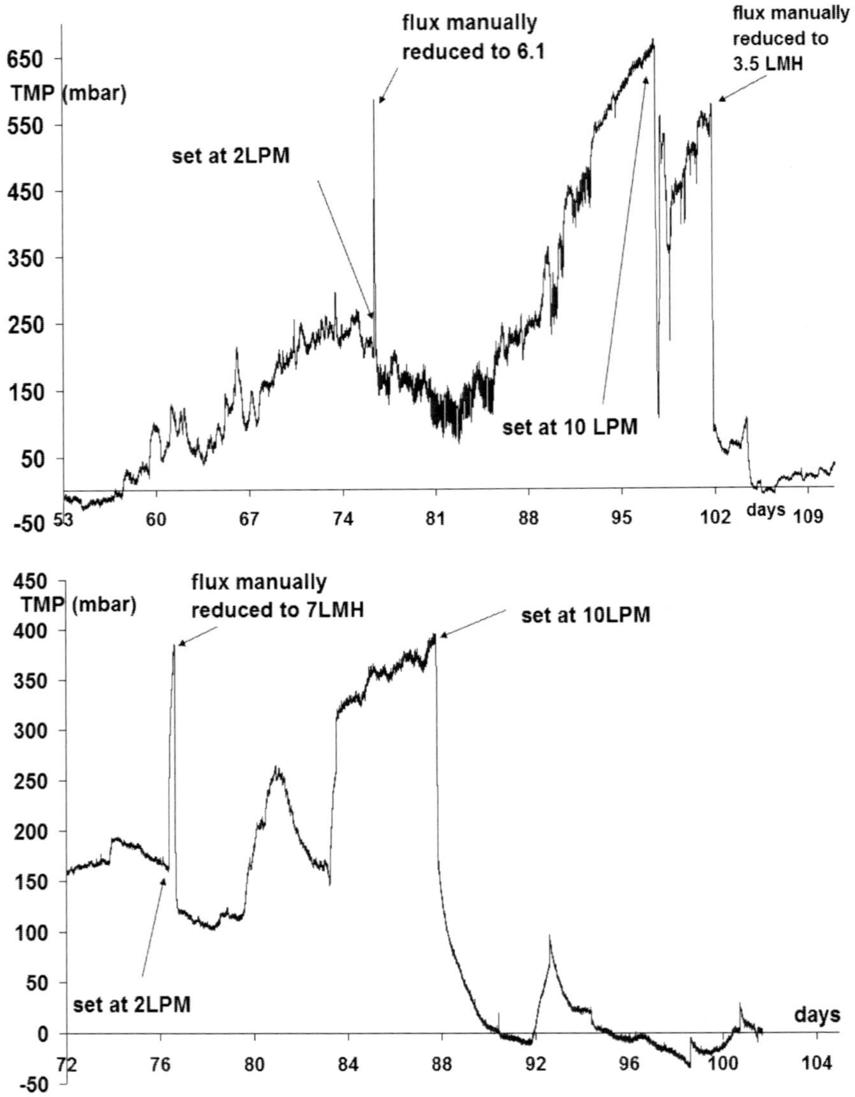

Figure 19. Evolution with time of the TMP in SAMBR1 (top) and SAMBR2 (bottom).

Therefore, it can be stated that in this study low sparging rates even for a short period of time lead to irreversible fouling because of internal fouling of the membrane pores. This is in contradiction with Vyrides and Stuckey (2009) who observed no significant increase in TMP or decrease in flux when operating the biogas pump in an intermittent mode (5 min OFF - 10 min ON) for 120 days. In our study, however, low sparging rates were likely to increase the thickness of the concentration polarization layer because of less turbulence, and this probably led to the aggregation of humic substances in the boundary layer. Humic acids can adsorb to particulate matter in the cake layer which results in a more compact and dense cake layer. In addition, the coagulation with calcium may have been enhanced in the concentration polarization layer where the concentration of humic acids is greater, resulting in the formation of aggregates on the membrane. This can lead to a thicker cake layer because calcium facilitates humic acids adsorption to particulate matter and membrane surfaces. Nghiem et al. (2006) also observed a greater TOC rejection associated with an increase in calcium concentration. Moreover, the highest TOC rejection coincided with the highest level of fouling which is consistent with our observations. Eventually on day 97, the sparging rate in SAMBR1 was also set at 8-10 LPM but the effect on the permeate COD was not immediate. In contrast, the effect of increasing sparging rate on the TMP was immediate on day 97: it suddenly dropped to 100 mbar, but unfortunately, the sparging rate could not be maintained at 10LPM due to foaming problems. Consequently, the TMP ended up at 550 mbar and eventually the flux was reduced manually by decreasing the speed of the recycle pump so that the flux was 3.5 LMH. As soon as the flux was lowered, the TMP went back to low values on day 101. It was only on day 104 that the permeate SCOD also rose as in SAMBR2 to values around 440 mg COD/L. Thus seven days (day 104-97) were necessary to see the effect of sparging rate increase in SAMBR1 that kept the low sparging rate mode for 21 days (day 97-76), while it took only 3 (day 90-87) days in SAMBR2 that kept the low sparging rate mode for 11 days (day 87-76).

The effect of the low sparging rate on the flux was immediate and detrimental. As can be seen from Figure 13 for SAMBR1 and from Figure 17 for SAMBR2, the flux drop was 4.8 (8.3 to 3.5LMH) and 3 LMH (10 to 7LMH) for SAMBR 1 and 2, respectively. Thus the effect of the low sparging rate was to block the pores while the partially clogged pores could still let the solutes pass. As a result of the pore blocking mechanisms, fewer molecules could pass which caused a significant increase in the bulk SCOD of both SAMBRs. Furthermore, the increase of the sparging rate to 10LPM did not allow the initial flux to recover showing that irrecoverable fouling occurred in both SAMBRs. The fouling was, however, worse for SAMBR1 because of the greater degree of compaction due to the longer period at low sparging rate, and also because its MLTSS was 5 times greater than that of SAMBR2 thus there was more material available in the bulk liquid to build a cake.

Therefore, this experiment performed in duplicate in separate SAMBRs showed similar results; when the sparging was set at a low rate (2 LPM), no significant change was noticed in the permeate COD. This might be due to longer periods required to build up a cake layer under slow TMP increase, and thereby obtain higher COD rejection and lower permeate COD. It is possible, however, that as the TMP increases, the compaction of the cake would be such that the pores of the biofilm will be so narrow that lower permeate COD will be obtained. Although the permeate SCOD did not decrease significantly, the membrane rejection started to increase significantly from day 80 onwards (Figure 20). Thus the effect of lowering the sparging rate was not a decrease of permeate COD, but instead an increase in membrane rejection due to enhanced concentration polarization, keeping at the same time a constant permeate SCOD. Choo and Lee (1996b) also concluded that cake compaction over time had more impact on the SCOD rejection by the membrane. This sounds paradoxical, but can be explained by the very wide range of molecular weight organics present in the bulk; while the lower MW can still pass through the biofilm and the membrane and are responsible for the constant permeate SCOD, the higher MW cannot pass through the pores and remain in the bulk which causes the membrane rejection to increase. These results are consistent with Choo and Lee (1996b) who, after the initial fouling of the membrane, observed a decrease in the SCOD in the bulk and the permeate due to the inoculum acclimatisation, but afterwards the permeate SCOD remained relatively constant for the rest of the experiment and did not decrease any further although the flux was decreasing due to internal fouling and cake layer compaction. This suggests that there is a limit below which the SCOD concentration in the permeate cannot be because the low MW recalcitrants which constitute the SCOD will still be able to pass through the membrane pores.

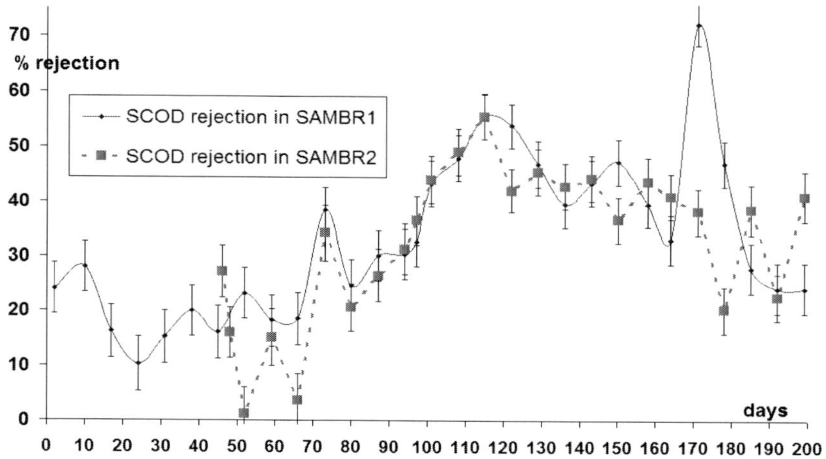

Figure 20. Evolution with time of the SCOD rejection by the membrane in SAMBR1 and SAMBR2.

On the other hand, when the sparging rate was increased to 8-10 LPM, the permeate COD increased typically from 360 to 440 mg/L due to a better scouring of the membrane, which resulted in a thiner concentration polarization layer and a better diffusion through the biofilm. Moreover, this increase in COD was faster for SAMBR2 that kept the low sparging mode for 11 days indicating that part of the cake layer can be easily scoured as indicated by the TMP that went back to zero after the high sparging rate was applied. In contrast, in SAMBR1 the low sparging rate mode was kept for 21 days, and due to the better compaction of the cake, the removal of the reversible fouling was impossible or much slower as indicated by a persistently high TMP, and as a result the effect of a higher sparging rate took longer to be noted. This increase in permeate SCOD was, however, not correlated with a reduction in membrane rejection in any of the SAMBRs shown in Figure 20.

Li et al. (2003) also concluded that the removal of the cake strongly depend on its age; if the cake existed for a short period of time, the lift forces due to surface shear can break some of the enmeshed bonding between the bacteria, allowing for the removal of the cake in the form of flocs. On the other hand, if the cake has been built up for a longer period, the possibility of breaking the enmeshment is reduced. The aggregation of the cake will also depend on ionic strength: at high ionic strength, which is likely to be the case in landfill leachate, the electrical double layer of bacteria is compressed resulting in more interaction and binding with molecules which allows a dense biofilm. Nevertheless, the change in COD (about 80 mg/L) was relatively small compared to the total COD concentration: this is a 20% increase in permeate SCOD by changing the sparging rate from 2 to 8-10 LPM.

3.4. Performance of the AMBR

3.4.1. Effect of the HRT on Nitrification

The sequential oxidation of NH_4^+ to NO_3^- involves autotrophic NH_3 oxidisers (AOB = ammonia-oxidising bacteria) and autotrophic NO_2^- oxidisers (NOB = nitrite oxidising bacteria). In addition, heterotrophic bacteria can oxidise reduced forms of organic N to NO_3^- (Prosser, 2007). Ammonia-nitrogen in the permeate of SAMBR2 varied between 45 and 215 mg/L. This concentration depends on the hydrolysis efficiency of proteins in the HR which in turn varied according to the VS retention time and the HRT of the HR. Moreover, the protein (only a few measurements were taken) concentration decreased whereas the ammonia concentration increased when passing through the SAMBR, which means that the hydrolysis of proteins was continuing in the SAMBR. Protein determination on days 115, 143, 171 and 199 gave a concentration circa 100 mg/L in the permeate of SAMBR 1 and 2, whereas it was around 200 mg/L in the leachate from the HR. Surprisingly, the concentration in the AMBR permeate was still around 50 mg/L although proteins are known to be degraded readily. It was found in the literature that the determination of proteins using the direct Lowry procedure can suffer from interference due to other chemicals such as tris, ammonium sulfate, EDTA, sucrose, citrate, amino acid and peptide buffers, and phenols. The procedure with protein precipitation, which uses deoxycholate and trichloroacetic acid, eliminates all these interferences with the exception of phenols (Bensadoun and Weinstein, 1976). As the procedure with protein precipitation was not used in this work, it is thought that the remaining 100 mg/L in the SAMBR permeate as well as the 50 mg/L in the AMBR permeate were recalcitrant phenolic or proteinaceous compounds interfering with the protein analysis. The 50% degradation of these phenolic compounds in the AMBR shows that further stabilization was taking place in the AMBR which supports the polishing character of the AMBR demonstrated by the COD removal. Figure 21 shows the evolution of inorganic nitrogen in the AMBR. Because the inoculum used in this study came from a dye wastewater plant, it was assumed that it did not contain any nitrifiers. As a result, ammonia-nitrogen was initially not converted to nitrite or nitrate. Ammonia oxidisers may also have been inhibited by undissociated ammonia (NH_3) which was in the range 14 - 23 mg NH_3/L between days 136 and 146. Anthonisen et al. (1976) have observed that free ammonia can inhibit ammonia oxidation to nitrite by *Nitrosomonas* and nitrite oxidation to nitrate by *Nitrobacter* or *Nitrospira* in the range 10-150 and 0.1-1 mg NH_3/L, respectively. The temporary nitrite build-up may be explained by the inhibition of nitrite oxidisers due to the free ammonia ranging from 0.1 to 0.4 between days 146 and 167. Inhibition of nitrifying organisms by free nitrous acid (HNO_2) is unlikely to have occurred as the concentration remained in the range 0.00084-0.0052 mg HNO_2/L, which is far below the inhibitory range of 0.22 to 2.8 mg/L reported by Anthonisen et al. (1976). The growth of *Nitrobacter* (or *Nitrospira*) was confirmed by the slow decrease in nitrite which was correlated with a slow increase in nitrate.

From day 171 onwards, on average 97.7% of the NH_4^+-N was converted to nitrite and nitrate in the AMBR at a maximum nitrogen loading rate of 0.18 g NH_4^+-N /L.day and a HRT of 0.5 day. According to Figure 21, the evolution of nitrogen is the one observed in batch mode

(Anthonisen et al., 1976) where the ammonia is sequentially converted to nitrite and then to nitrate. In full scale continuous wastewater treatment plant nitrite is seldom found in the reactors because the conversion of NO_2^- to NO_3^- is much faster than the conversion of $NH_4^+$to NO_2^-. Therefore, as long as the NOB species are present in the digester running continuously nitrite is almost never found. In this study, nitrite was found because the inoculum did not contain any nitrifiers, and the presence of a membrane kept all the species in the reactor and thereby resulted in a sequential reaction as if the AMBR was batch fed. Regarding the effect of the HRT, it can be said that the growth of AOB and NOB was not impeded by the successive HRT shocks until the shock to 0.99 day where the nitrite concentration did not reach zero which indicates that this HRT could overload the NOB such as *Nitrobacter*. However, previous work has shown that floc size for ammonia degradation is also important (Wittebolle et al., 2008) because a close distance between the nitrite producers and nitrite oxidisers is required to obtain complete nitrification. Wittebolle et al. (2008) have compared nitrification in a sequencing batch reactor and a MBR, and found that residual nitrite in the MBR was correlated with smaller flocs. They measured flocs of 1 to 40 microns in the MBR versus 300 microns in the sequencing batch reactor. Because of the shear effect in the AMBR, it is possible that the residual nitrite was due to a lack of syntrophic association between AOB and NOB. In our study, particle size distribution analysis was carried out in the AMBR, but particles of 1000 microns were detected which could be inorganic precipitates, thus rendering the comparison with data in the literature difficult. In another study (Shin et al., 2005), a temporary nitrite build-up in a MBR was associated with a poor oxygen transfer in the dead space of the MBR. The authors also reported that nitrification was better accomplished with pH control with HCl than without pH control. In this study, the SCOD fed to the AMBR was relatively low (400-600 mg/L) which promoted the growth of autotrophic bacteria. Because of the low organic content, high DO (1.6 mg/L) and good mixing (1.4 LPM) optimal conditions were met for the growth and retention of autotrophic AOB in the AMBR at a HRT as low as 0.37 day. In contrast, Chen et al. (2008a) and Im et al. (2001) could not maintain nitrification at 1.5 and 2.7 days HRT, respectively, because the COD concentration in the feed to the aerobic step sharply increased. Jokela et al. (Jokela et al., 2002) also observed that nitrification efficiency dropped to below 20% when the COD concentration suddenly increased at 1.4 d HRT. As a result, heterotrophs competed for oxygen with the autotrophs leading to a decrease in nitrification activity. The AMBR permeate containing circa 50 mg/L of nitrite and the same concentration of nitrate was recycled to the HR where no nitrite or nitrate was detected. This was confirmed by the NH_4^+-N and TN analysis which gave close concentrations; typically the TN was 10-15 mg/L greater than the NH_4^+-N concentration in the HR effluent. The discrepancy was organic nitrogen such as proteins, peptides, nucleic acids or urea; this strongly suggests that denitrification took place in the HR. Another clue is obtained by comparing the TN concentration of the AMBR effluent and the HR effluent; the latter was on average 67% lower than the former which indicates substantial N removal. At this point it is not clear whether Anammox species were responsible for the conversion of nitrite or nitrate to gaseous nitrogen using ammonia that is the end product of protein degradation from the OFMSW feedstock fed to the HR. It is known that NOB can convert nitrite to nitrate if dissolved oxygen is present, and that heterotrophs will consume all the oxygen present in the HR before the NOB could consume the oxygen. However, as the HR was frequently opened for feeding and because of technical problems, it is possible that there was enough DO for NOB to grow and coexist with heterotrophs. Figure 8 shows how often the HR was opened for technical problems, and it can be seen that after day 160 there was no methane in the headspace. As the DO was not measured in the HR, it is not clear whether NOB could have grown in the HR, but denitrifiers on the other hand were likely to grow because of nitrate from the AMBR and the carbon sources available. Moreover, the TN analysis also revealed that between 7 and 35% of the TN in the permeate of the AMBR was organic N and that organic N was slowly increasing over time. Hence, heterotrophs could very likely have coexisted in the AMBR using organic N for growth and recalcitrants as a sole carbon source, which may explain why the AMBR acts as a COD-polishing step. On one hand there was organic N as a nitrogen source for the heterotrophs in the AMBR, but on the other hand their increase suggests that these were recalcitrant molecules containing amine groups.

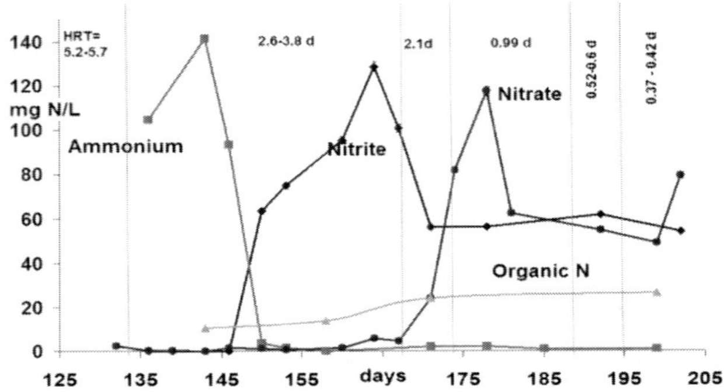

Figure 21. Concentration over time of the organic and inorganic nitrogen in the AMBR permeate. Reprinted from (Trzcinski and Stuckey, 2009b).

3.4.2. Effect of the HRT on the TSS, VSS, FSS, pH and Flux

The AMBR was started up at a TSS, VSS and FSS of 3, 2.3 and 0.7 g/L, respectively, and was fed with the stabilized leachate, i.e., the permeate of SAMBR2. Hence the feed contained no solids in suspension. Figure 22 shows that the VSS/TSS decreased linearly with time because the biomass was either settling out, attaching to the reactor's panels, or was forming a biofilm on the membrane.

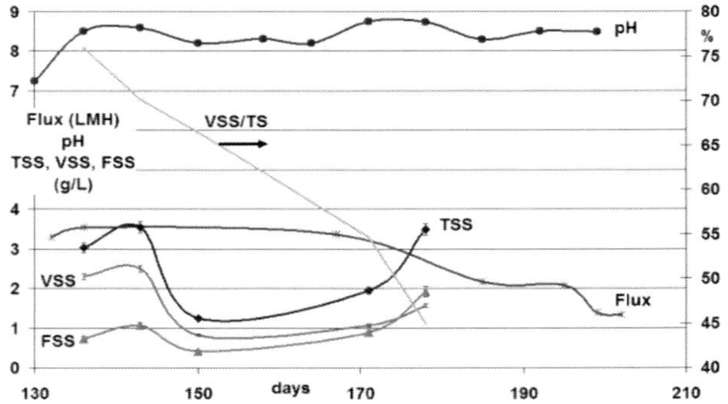

Figure 22. Evolution with time of the pH, flux, TSS, VSS, FSS (left axis) and the VSS/TSS ratio (right axis) in the AMBR.

The pH started at 7.25 but then plateaued at circa 8.5. This is due to the carbon dioxide being stripped out of the liquid phase due to air sparging while the reactor was open to the atmosphere. The feed to the AMBR was saturated in CO_2 because it was used in SAMBR2 to recirculate the biogas at the bottom of SAMBR2 for mixing and scouring purposes. The concentration is CO_2 is set by Henry's law:

$$[CO_{2(aq)}] = \frac{pCO_2}{K_{CO2}} \quad (10)$$

where pCO_2 is the partial pressure in atmosphere and K_{CO2} is Henry constant. In SAMBR2, as the biogas contains about 20% CO_2, it can be said that the partial pressure was 0.2 atm and the CO_2 concentration in the liquid is therefore $5.24.10^{-3}$ M using Henry's law described in Equation 10. In the AMBR which is an open system, the partial pressure of CO_2(g) is relatively constant at 0.000355 atm (=355 ppm). Therefore, CO_2(aq) is equal to $1.38.10^{-5}$ M. This is almost 3 orders of magnitude less than in the SAMBR. In other words, as soon as the feed enters the AMBR the carbon dioxide is stripped out to reach a new equilibrium according to Henry's law. As a result of CO_2 stripping the equilibrium of carbonic acid is changed and there is less dissolved carbon dioxide available for carbonic acid formation:

$$CO_{2(aq)} + H_2O \rightarrow H_2CO_{3(aq)} \qquad K_1 = 1.18 \cdot 10^{-3} \quad (11)$$

There is, therefore, less dissociation into protons according to the carbonic acid equilibrium, which caused the pH to rise.

$$H_2CO_{3(aq)} \rightarrow H^+ + HCO_3^-{}_{(aq)} \qquad K_2 = 8.38 \cdot 10^{-4} \quad (12)$$

$$HCO_3^-{}_{(aq)} \rightarrow H^+ + CO_3^{2-}{}_{(aq)} \qquad K_3 = 4.2 \cdot 10^{-11} \quad (13)$$

Regarding the alkalinity, it can be defined as:

$$A=[HCO_3^-]+2[CO_3^{2-}]+[B(OH)_4^-]+[OH^-]+3[PO_4^{3-}]+[HPO_4^{2-}]+[SiO(OH)_3^-]-[H^+]-[HSO_4^-]-[HF] \quad (14)$$

but this can often be approximated by:

$$A= [HCO_3^-] + 2 [CO_3^{2-}] + [OH^-] + [H^+] \quad (15)$$

because phosphates and silicate are negligible, and at pH = 8.5, $[HSO_4^-]$ and [HF] are also negligible. As carbon dioxide is stripped out of the system, the pH increases, but the removal of carbon dioxide does not modify the alkalinity of the system. At pH 8.5, the ion HCO_3^- predominates over H_2CO_3 and CO_3^{2-}. Thus when carbon dioxide is removed from the system, the following equilibrium is shifted to the left:

$$CO_{2(aq)} + H_2O \leftarrow HCO_3^- + H^+ \quad (16)$$

The alkalinity is not changed because the net reaction removes the same number of equivalents of negatively contributing species (H^+) as positively contributing species (HCO_3^-). The precipitation of calcium carbonate, however, has a strong influence on the alkalinity. This is because carbonate is composed of $CaCO_3$ and its precipitation will remove Ca^{2+} and CO_3^{2-} from the solution. The calcium ion will not influence alkalinity, but CO_3^{2-} will decrease the alkalinity by 2 times the molar concentration of CO_3^{2-}.

The precipitation of calcium carbonate was observed in our system, and the decrease of alkalinity was confirmed by titrating the effluent from SAMBR2 and AMBR which were found to contain 2250 and 750 mg eq $CaCO_3$/l, respectively (Figure 23). Borzacconi et al. (1999) have also observed that calcium ions can precipitate as $CaCO_3$ at a pH around 8. The maximum concentration of calcium before precipitation can be predicted using the solubility product assuming that the interference of other ions can be neglected. The equilibrium is given by the equation:

$$CaCO_{3(s)} \rightarrow Ca^{2+} + CO_3^{2-} \quad (17)$$

where the solubility product varies from $K_{sp} = 3.7 \cdot 10^{-9}$ to $8.7 \cdot 10^{-9}$ at 25°C, depending upon the data source (Seely, 2008; Weast, 1963). However, this equation must be taken along with the equilibrium of carbon dioxide with atmospheric carbon dioxide and the equilibrium of bicarbonate and carbonic acid given above.

For the AMBR, it is assumed that the partial pressure of carbon dioxide is $3.5 \cdot 10^{-4}$ atmospheres and that the pH is 8.5. Therefore, according to Henry's equation, $CO_{2(aq)}$ is equal to $1.38 \cdot 10^{-5}$ M.

This results in a concentration of H_2CO_3 equal to $1.6.10^{-8}$ moles per litre taking into account the dissociation constants given in equations 11, 12 and 13. When H_2CO_3 is known, the remaining three equations together with (Perrin, 1982):

$$H_2O \rightarrow H^+ + OH^- \quad K = 10^{-14.167} \text{ at } 20°C \qquad (18)$$

makes it possible to solve simultaneously for the unknown concentrations. If $4.47 \cdot 10^{-9}$ is taken for K_{sp}, the maximum solubility of Ca^{2+} in the AMBR is only 3.1 mg/L at pH 8.5. Furthermore, this value shows a quadratic dependence with the concentration of H^+ so that the maximum Ca^{2+} concentration is 100 times greater at pH 7 than at pH 8.

In SAMBR2 at 35°C, the pH was 7 and the partial pressure of carbon dioxide was 0.2 atm (maximum 20% CO_2 in the headspace), and the CO_2 concentration in the liquid phase is therefore $5.24 \cdot 10^{-3}$ M. The solubility of calcium at pH 7 is circa 2,850 mg/L, but because of the high partial pressure of carbon dioxide the solubility in SAMBR2 was only 5.1 mg/L. Thus precipitation can occur in the SAMBR as well as in the AMBR if the calcium concentration is greater than the calculated values. If the pH increases to 7.3 due to high ammonia concentrations for instance, the maximum solubility of calcium become 1.3 mg/L, while if acidification to pH 6.8 takes place due to high VFA concentrations, it becomes 12.7 mg/L. Figure 22 also shows that the flux in the AMBR was decreasing over time. Even though the TMP could not be recorded because only 2 transducers were available for the SAMBRs, it is very likely that calcium carbonate precipitating on the membrane was causing the flux to drop. Visual inspection of the membrane indicated a hard, off-white, chalky deposit similar to limescale, and contact with a drop of HCl provoked an immediate fizzing. A precipitate was also found in the permeate tubing, and a sample was taken and washed with plenty of water and then dissolved in deionized water with HCl. It was then filtered through a 0.2 micron filter and analysed on a Dionex Ion Chromatograph that confirmed that the main element was calcium. In the HR, if a pH of 6 and a CO_2 partial pressure of 0.3 atm are considered, the maximum calcium solubility is 954 mg/L, indicating that precipitation is less likely to occur in the HR.

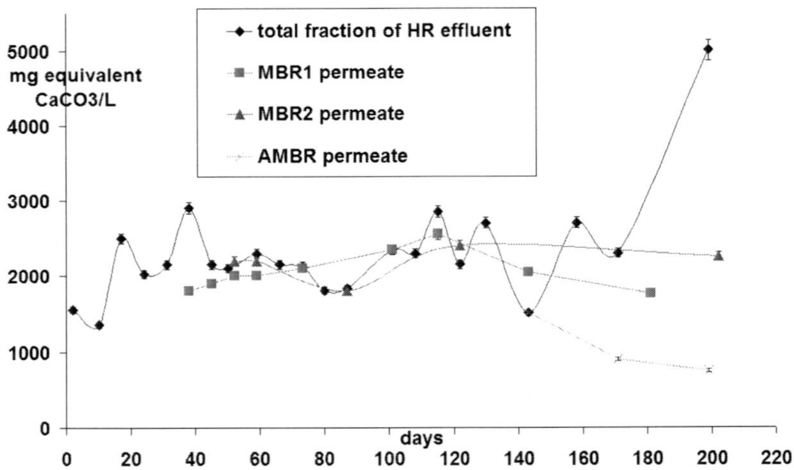

Figure 23. Evolution of the alkalinity as mg equivalent $CaCO_3$ per litre in the effluent of HR and the permeate of SAMBR 1, SAMBR2 and AMBR.

3.4.3. Scanning Electron Microscope - Energy Dispersive X-Ray (SEM-EDX)

Scanning Electron Microscope (SEM) was used to investigate the morphology and structure of the inorganic precipitate found on the membrane in the AMBR. This technique was coupled to Energy-Dispersive X-rays (EDX) in order to determine the chemical composition of the sample. The EDX spectra confirmed that the crystals formed contained calcium, oxygen and phosphorus. The carbon peak observed was likely to be due to the sample carbon coating, but it could have come from the membrane material itself (polyethylene). Some oxygen can also be attributed to dissolved organic matter although it is noteworthy that it can also be from the membrane polymer. Several random points on the samples were analyzed by EDX and all the spectra exhibited calcium, oxygen and phosphorus as the main elements. It was found that calcium could precipitate in two different phases; the flat background precipitate (Figure 24), or as long and flat needles (Figure 25) of about 70 microns pointing upwards. The former phase contained small amounts of manganese, iron, magnesium, manganese, silicon, aluminium, sulphur, chloride and sodium (<1% by weight) whereas the needles contained relatively few impurities, which indicates that the needles are a purer form of precipitate than the background (Table 10). Moreover, the needle-shaped precipitate was characterized by significantly higher amounts of phosphorus compared to the background. Unfortunately, as the samples were coated with carbon, the carbon percentage was not available. Thus the comparison of the weight percentages from Table 10 with known minerals is difficult. Nonetheless, it is very likely that the background precipitate was comprised of calcium carbonate containing many impurities.

Figure 24. General view of the precipitate on the membrane in the AMBR (left), and elemental chromatogram of the background (right). Reprinted from (Trzcinski and Stuckey, 2016).

Regarding the needles, calcite was not considered due to the high phosphorus content. Several authors have reported that the presence of phosphates inhibits calcite growth due to the adsorption of phosphates on the calcite surface, which enables the formation of calcium phosphates (Lin and Singer, 2005; Plant and House, 2002).

Figure 25. Zoom of the needle-shaped precipitate on the membrane in the AMBR (left), and its elemental chromatogram (right). Reprinted from (Trzcinski and Stuckey, 2016).

Interestingly, it was observed in the SEM picture (Figure 25) that indeed the needles seem to form on the background of calcite, and grow perpendicular to the background. The high amount of phosphorus suggests that it could be a calcium phosphate such as calcium hydrogen phosphate (or brushite $CaHPO_4.2H_2O$, $K_{sp} = 10^{-6.69}$ (Montastruc et al., 2003)), dicalcium phosphate anhydrate ($CaHPO_4$, $K_{sp} = 10^{-6.9}$ (Montastruc et al., 2003)), amorphous calcium phosphate ($Ca_3(PO_4)_2$, $K_{sp} = 10^{-24}$ (Mamais et al., 1994)), octacalcium phosphate ($Ca_4H(PO_4)_3.2.5H_2O$, $K_{sp} = 10^{-49.6}$ (Montastruc et al., 2003)), hydroxyapatite ($Ca_5(PO_4)_3OH$, $K_{sp} = 10^{-46.8}$ (Barat et al., 2008)) or carbonated hydroxyapatite ($Ca_{10}(PO_4)_6(CO_3)_x(OH)_{2-2x}$, $K_{sp} < 1 \cdot 10^{-100}$ (Ito et al., 1997)). However, calcium phosphate precipitation follows the Ostwald rule, which states that the least thermodynamically stable phase is the first one formed, which works as a precursor of the most stable phase (Montastruc et al., 2003). In this case, carbonated hydroxyapatite would be the least thermodynamically stable phase forming on the calcite background as a precursor for a more stable calcium phosphate. Furthermore, the same authors (Montastruc et al., 2003) found that both dicalcium phosphate dihydrate and amorphous calcium phosphate can crystallize at a pH lower than 7.3, and only amorphous calcium phosphate at a pH higher than 7.3. Therefore, the more stable phase in this study would either be dicalcium phosphate dihydrate or amorphous calcium phosphate.

Table 10. Weight percentages of several points of the background and the needle precipitates

Background	Element										
	O	Ca	P	Al	Si	S	Fe	Mg	Cl	Na	Mn
Spectrum 1	46.7	47.98	2.32	0.64	1.09	0.28	0.99	0	0	0	0
Spectrum 2	38.95	55.23	2.08	0.3	1.06	0.28	1.49	0.31	0.29	0	0
Spectrum 3	45.03	50.3	3.43	0	0	0	0.66	0.57	0	0	0
Spectrum 4	55.93	33.55	6.66	0.26	0	0.17	0.82	0.77	0	0.82	1.02
Needles	Element										
Spectrum 5	55.79	26.19	17.22	0.22	0.58	0	0	0	0	0	0
Spectrum 6	30.99	59.04	9.97	0	0	0	0	0	0	0	0
Spectrum 7	64.04	18.78	17.18	0	0	0	0	0	0	0	0

Le Corre et al. (2005) and Doyle and Parsons (2002) observed that calcium ion can affect struvite crystal growth and leads to the formation of an amorphous calcium phosphate. Therefore, it is very likely that struvite precipitation did not occur because the available phosphates precipitated with calcium on the membrane keeping the solution under saturated with respect to struvite. From all these considerations and from the light shed by other authors, it seems very likely that the background was amorphous calcium carbonate and the presence of phosphates inhibited calcite growth due to the adsorption of phosphates on the calcite surface, which enabled the formation of calcium phosphates. The low level of impurities in these needles suggest that they have a more crystalline structure which is why hydroxyapatite is the most likely. Chesters (2009) obtained very similar SEM pictures with flat needles which he assumed to be hydroxyapatite crystals. In the same paper, the authors report the use of a new anti-scalant (Genesys PHO: 2-5 mg/L) to control calcium phosphate scale by inhibition mechanisms so that the solubility of calcium phosphate is raised 150 times.

3.4.4. Particle Size Distribution

It is well known that particles having the same size as the membrane pore are particularly likely to cause fouling. The particle size of the leachate and the contents of the SAMBR and AMBR were measured and compared in Figure 26. It can be observed that the leachate is very heterogeneous as it contains particles ranging from 0.3 microns to 1 mm. As expected, SAMBR1 and SAMBR2 also contains particles with a wide range of sizes including colloids below 0.4 microns which are presumably the particles responsible for pore blocking. The distribution of the leachate and SAMBRs mixed liquor were bimodal with a peak at circa 1 micron and the other at 10 microns, and a high proportion (90%) was below 48, 83 and 30 microns for the HR effluent, SAMBR1 and SAMBR2 mixed liquor, respectively. No straightforward comparison could be made between the leachate and the SAMBR bulks because of the heterogeneous character of the leachate that contained bacterial flocs as well as lignocellulosic fibres. The comparison between the volume percentages are also difficult as the accuracy of the MasterSizer 2000 is within 5% and the peaks on Figure 26 are all smaller than 5%. However, the distribution of the AMBR particles is such that 90% was below 632 microns with a mean diameter d(4,3) equal to 405 microns.

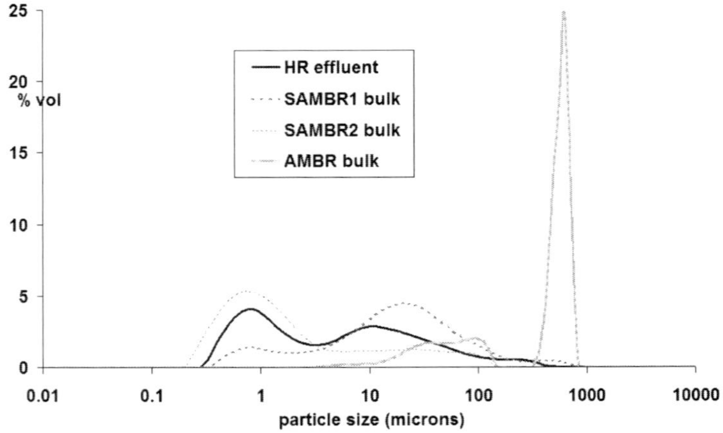

Figure 26. Particle size distribution of the leachate from the HR and the mixed sludge from SAMBR1, SAMBR2 and the AMBR.

3.5. Salts

3.5.1. Introduction

The salts ions are Na^+, K^+, Mg^{2+}, Ca^{2+}, Cl^-, PO_4^{3-}, SO_4^{2-}, and it is well known that high salt levels can cause bacterial cells to dehydrate due to osmotic pressure and that Na^+ is more toxic that any other salt on a molar basis. It is also known that the cation determines the toxicity of salts (McCarty and McKinney, 1961). Sodium, potassium, magnesium, calcium, phosphate and sulphate are present in the influent of anaerobic digesters and they are all required for microbial growth. Moderate concentrations stimulate microbial growth, but excessive amounts can cause inhibition or toxicity (Soto et al., 1993). Figure 27 shows the evolution with time of the ions. Sulfate was omitted because its concentration remained below 10 mg/L in all reactor effluents throughout the 200 days. These results are discussed separately in the following sections.

3.5.2. Sodium

Sodium is essential for methanogens and concentrations in the range 100-200 mg/L were reported beneficial for the growth of mesophilic anaerobes (McCarty, 1964) because of its role in the formation of adenosine triphosphate or in the oxidation of NADH (Feijoo et al., 1995). Sodium concentrations ranging from 3,500 to 5,500 mg/L can be moderately inhibitory and 8,000 mg/L can be strongly inhibitory to methanogens at mesophilic temperature depending on the adaptation period, antagonistic/synergistic effect, substrate and reactor configuration (Kugelman and McCarty, 1964). He et al. (He et al., 2006) report that 3.5-5.5 g/L and 2.5-4 g/L are the concentrations of sodium and potassium ions, respectively, that are optimal for AD. Shin et al. (Shin et al., 1995) showed that 2 to 10 g/L moderately inhibited the methanogenic activity, but concentrations above 10 g/L strongly inhibits. Although 8 g/L of Na^+ and 12 g/L of K^+ causes severe toxicity in the treatment of wastewaters, methanogens have been reported to acclimate to high salts stress, such as 21.5 g/L Na^+ and 53 g/L Na^+. Furthermore, He et al. (2006) observed that 50 g/L Na^+ inhibited acidogenesis, but surprisingly had the highest amount of substrate hydrolysed because of the highest α-amylase activity at that concentration of Na^+; the bacteria recovered at a concentration of 25 g/L.

They also observed that denitrification was optimal at pH 7 and that 50 g/L suppressed denitrification. In another study, sodium was found to be more toxic to propionic acid-utilizing microorganisms than acetate-utilizing and lignocellulose-degrading ones (Liu and Boone, 1991; Soto et al., 1993). The literature gives confusing data regarding the optimal and inhibitory levels of salts, and papers have focused more on inhibition studies in batch bottles rather than in continuous systems. Indeed, very little data is available for full-scale plants as ions are not systematically measured to assess the health of anaerobic digesters.

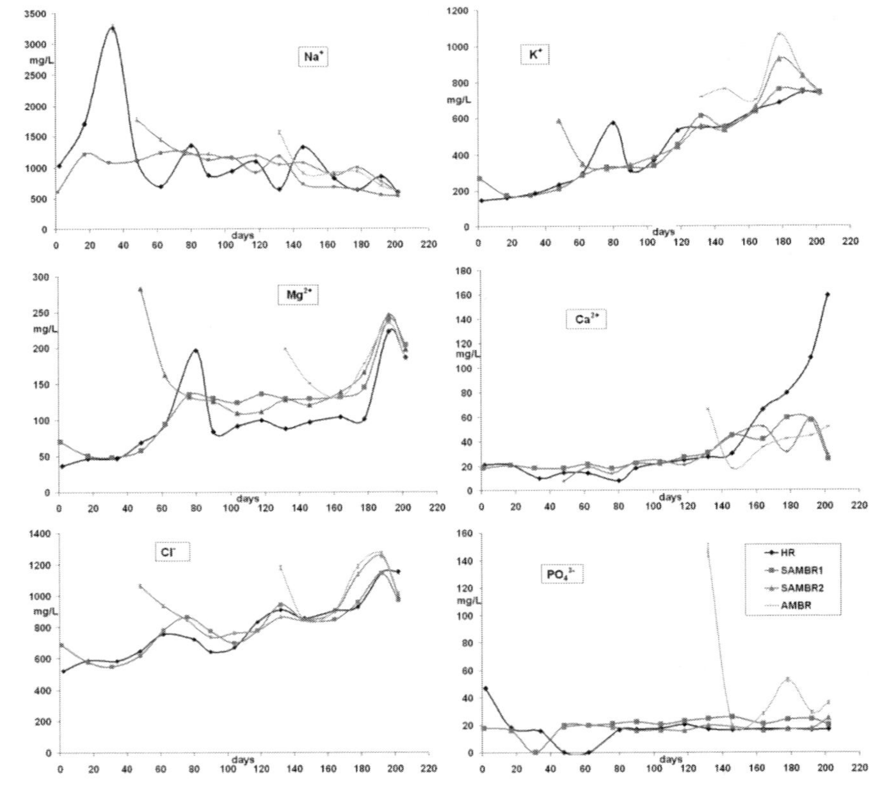

Figure 27. Concentration over time of Na^+, K^+, Mg^{2+}, Ca^{2+}, Cl^-, PO_4^{3-} in the HR effluent and the permeate of SAMBR1, SAMBR2 and AMBR. The error bars represent the standard deviation.

In this study, HR and SAMBR1 started with a Na^+ concentration of 1,034 and 612 mg/L, respectively, whereas SAMBR2 and AMBR started with a sodium concentration over 1.5 g/L due to the sodium present in the initial biomedium (Owen et al., 1979) that already contains 1373 mg Na^+/L. The concentration in the HR effluent and SAMBR1 permeate increased markedly to over 1000 and 3000 mg/L, respectively, in the first 40 days during which time the OLR was low at 0.5 g VS/L.day. There was an initial build-up of sodium due to the high solids destruction, and very little amount of digestate were taken out of the reactor. The levels in both reactors then stabilized around 1 g/L presumably due to sodium uptake for bacterial growth. When SAMBR2 and AMBR started up their levels levelled off also at circa 1 g/L suggesting a sodium uptake for bacterial growth, and this uptake was not renewed by new sodium ions from the feedstock as the OLR was increased and the solids destruction decreased. Also the sodium content of the OFMSW was low (2.3 ± 0.3 mg Na/g of dry matter). The limited solids destruction at higher OLRs can also explain why more solids were taken out on a daily basis and consequently, the sodium may have been washed out in the final digestate that still contained a bit of process liquid. However, it is thought that this hypothesis was not the main reason because if it was the case the concentration of all other ions would also have decreased over time and that was not the case. It can be seen that at the end of the run, the concentration was below 700 mg/L in all the reactor effluents meaning that the concentration was suboptimal according to He et al. (2006); Owen et al.'s biomedium (1979) contains circa 1,373 mg Na^+ due to $Na_2S.9H_2O$ and $NaHCO_3$. This decrease of sodium could have had a severe impact on the process; it could have impeded the growth of new bacteria and decreased the hydrolysis performance because of less hydrolytic enzymes produced as He et al. (2006) stated that high sodium levels promoted α-amylase activity. It could also cause dehydratation of cells resulting in their lysis, therefore, it is suggested to maintain the level of sodium between 1 and 2 g/L by addition of more kitchen waste that contains more sodium.

3.5.3. Potassium

Potassium is known to increase cell wall permeability by aiding the cellular transport of nutrients and providing cation balancing (Kayhanian and Rich, 1996). Low concentrations (less than 400 mg/L) can cause an enhancement in performance in both the mesophilic and thermophilic ranges (Chen et al., 2008b) while Kugelman and McCarty (1964) have observed that 0.15 M K^+ (=5.9 g/L) caused 50% inhibition of acetate-utilising methanogens. Sodium, magnesium and ammonium can also mitigate potassium toxicity, with sodium producing the best results. Figure 27 shows that potassium in all reactors built up steadily over 200 days independently of the OLR. However, the evolution in the HR effluent increased suddenly from 297 on day 62 to 572 mg K^+/L on day 80 when the OLR was set at 2 g VS/L.day. On the other hand, the evolution in SAMBR1 permeate was quite linear (R^2=0.91) suggesting that an equal amount of potassium was solubilised independently of the OLR or the HRT applied. This slow increase also indicates that the potassium uptake for bacterial growth is not as important as the sodium uptake. As a result, potassium is likely to build-up in processes and could become inhibitory after longer periods that the one in this study and should therefore be monitored closely. The final concentration in all the reactors effluent was close to the theoretical concentration in Owen et al.'s biomedium (1979) of 680 mg K^+/L, thus it can be stated the concentration was optimal in the process towards the end of the experiment.

3.5.4. Magnesium

The optimal Mg^{2+} concentration was reported to be 720 mg/L for the anaerobic bacterium *Methanosarcina thermophila* (Schmidt and Ahring, 1993). High concentrations have been shown to stimulate the production of single cells which is detrimental because these are more sensitive to lysis and this is an important factor in the loss of aceticlastic activity (Chen et al., 2008). This study has shown that magnesium does not reach levels higher than 300 mg/L at all the OLRs tested. Although an increase was observed in the HR and SAMBR1 effluent when the OLR was changed to

2 g VS/L.day, the magnesium level remained relatively stable at an OLR of 4 to 16 g VS/L.day between 100 and 150 mg/L, which is relatively close to the level of 215 mg Mg^{2+}/L in Owen et al.'s biomedium (1979). This indicates that the OFMSW contains enough magnesium in itself and that no extra source would be required for optimal concentration. The level increased suddenly in all reactor effluents after day 178 due to an artificial spike of trace elements based on Owen et al.'s biomedium (1979) in order to investigate the effect of trace metal addition, but no positive or negative effect were observed.

3.5.5. Calcium

Calcium is known to be essential for the growth of certain strains of methanogens. It is also important in the formation of microbial aggregates. However, excessive amounts lead to the precipitation of carbonate and phosphate which may result in the scaling of reactors and biomass and reduced specific methanogenic activity, loss of buffer capacity and essential nutrients for anaerobic degradation (Chen et al., 2008b). Regarding the toxicity, the literature provides confusing data. No inhibitory effect was observed at 7000 mg Ca^{2+}/L (Jackson-Moss et al., 1989), but Kugelman and McCarty (1964) reported that moderate inhibition can happen at concentration ranging from 2,500 to 4,000 mg/L and that the optimum concentration for methanogens was 200 mg/L. Concentrations greater than 300 mg/L were reported to be detrimental for biofilm formation (Hulshoff Pol et al., 1983). Mathiesen (1989) found that calcium and magnesium supplementation enhanced the methane production and foaming was avoided.

In this study, calcium concentration was low (about 20 mg/L) for the first 100 days until the OLR was set at 4 g VS/L.day on day 97. As the levels were very similar in HR effluent and SAMBR1 permeate, it seems that precipitation did not occur because the calcium concentration was too low. As a reference, Owen et al.'s biomedium (1979) contains 68 mg Ca^{2+}/L in the form of $CaCl_2.2H_2O$. Interestingly, the level started increasing very rapidly to 160 mg/L in the HR effluent but the concentration in the other reactors remained much lower at about 60 mg/l, and even decreased to 20 mg/L at the end of the run indicating precipitation of calcium carbonate in the SAMBR and AMBR. The calcium concentration in the HR seems to be directly proportional to the OLR applied suggesting that calcium salts are readily dissolved in the HR but it could be also due to the lower pH in the HR that dropped below 6.5 after day 180. The concentration reached 80 mg/L on day 178 but the increase after day 178 is also partly due to the artificial spike of trace elements.

3.5.6. Phosphate

PO_4^{3-} was found at a relatively low concentration in all reactors, except in the AMBR that started up at about 140 mg/L, presumably due to the level in the inoculum. The concentration in the AMBR dropped down to low levels (20 - 40) like in the other reactors. Phosphate is essential for bacterial growth because it is contained in DNA, but high levels are not recommended because it can precipitate as struvite ($NH_4MgPO_4.6H_2O$). Because phosphate concentration is usually lower than ammonia and magnesium in anaerobic digester, it will determine whether struvite precipitates or not.

3.6. Size Exclusion Chromatography (SEC)

3.6.1. Introduction

Figure 28 presents the calibration curves used for the SEC. The standards with molecular weights (MW) greater than 219.3 did not appear as a single peak, and could not be resolved using the column Aquagel 0H-40. An exponential curve was fitted to the calibration standard and the following equation was obtained ($R^2 = 0.98$):

$$MW\ (Da) = 4 \cdot 10^{10} \cdot \exp(-0.9851 \cdot t) \qquad (19)$$

where t is the elution time.

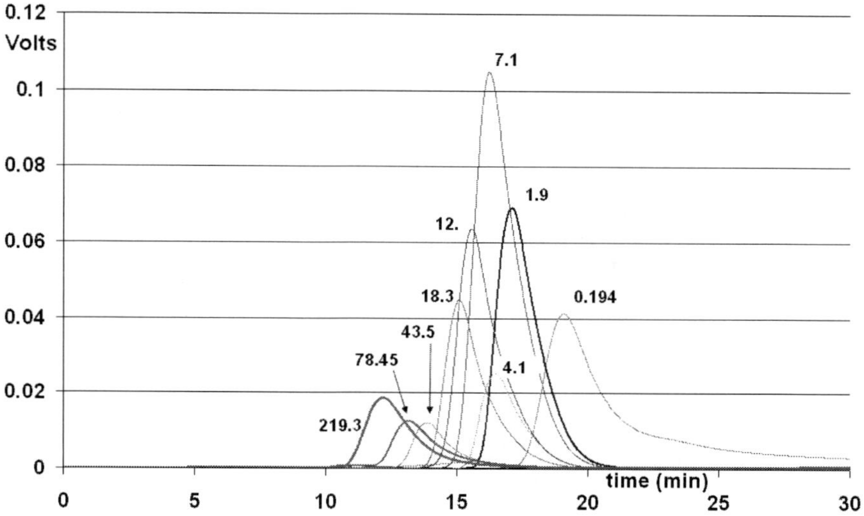

Figure 28. Calibration of the Size Exclusion Chromatography. The Y-axis is the absorbance of the PEG and PEO standards on the refractive index detector.

Samples of the HR effluent and the SAMBR and AMBR permeates were taken regularly during the continuous experiment in order to investigate the evolution with time of the molecular weight (MW) distribution in the reactor effluents. However, a first objective was to study the difference in MW between the different reactors at a specific moment. Figure 29, Panel A, shows the size exclusion chromatograms of the HR effluent, SAMBR1 bulk and permeate, SAMBR2 bulk and permeate and AMBR bulk and permeate on day 192. This day was deemed representative of the situation in the process because the SCOD in the SAMBR and AMBR was stable. All the chromatograms contain two main peaks: one in a region of high MW (circa 11.5 min) and a second sharper peak at low MW (around 17.5 min). A similar observation was made by Ortega-Clemente et al. (2008) and Sierra-Alvarez et al. (1990). The former team was treating pulp mill effluents in an anaerobic fluidized bed reactor, while the latter treated soda pulp liquors of pine, spruce and wheat straw in a UASB. They stated that the lignin fraction removed or bio transformed anaerobically corresponds to low MW lignin derivatives.

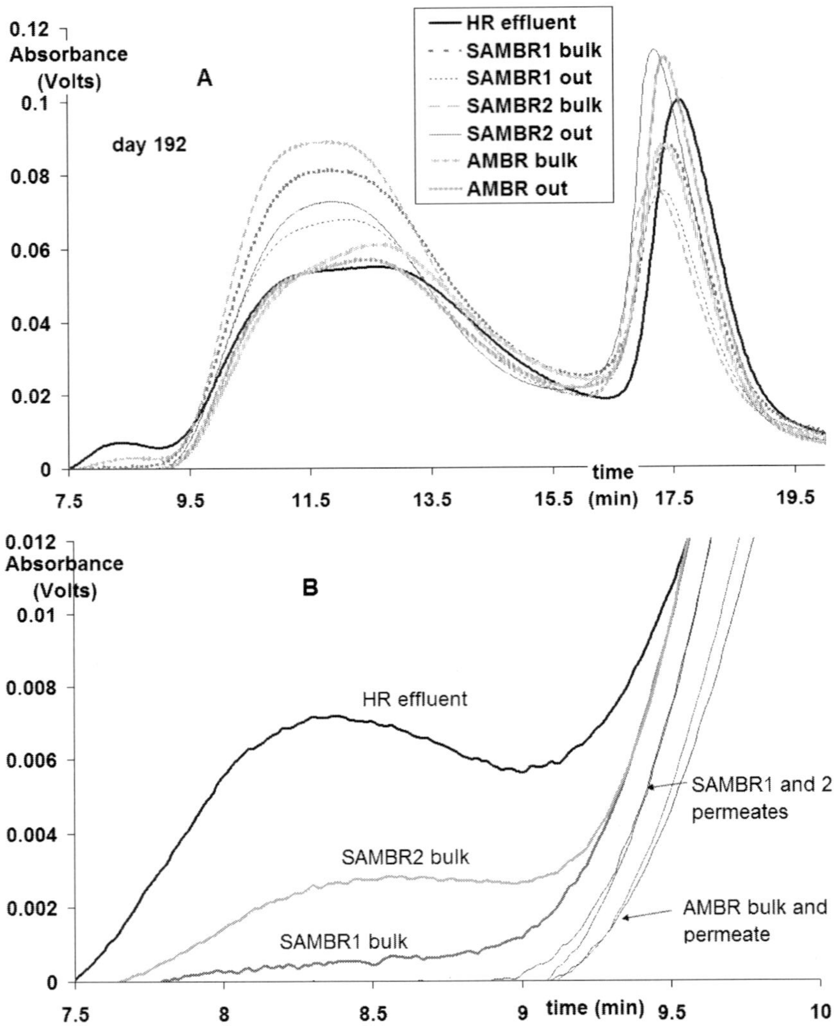

Figure 29. (A) Size exclusion chromatograms of the HR effluent, SAMBR1, SAMBR2 and AMBR bulks permeates on day 192. (B) Zoom of the curves before 10 minutes of elution time.

The absorbance on the UV detector set at 254 nm was zero before 7.5 minutes and after 20 minutes. Panel B shows a zoom of the very high MW ($t \leq 9$ min) in order to understand the fate of very high MW organics (MW >> 200kDa). Panel B shows clearly that the very high MW organics were dominant in the leachate fed to the SAMBRs; the bulk of the SAMBRs still contained some of these molecules but considerably less. Interestingly, their permeate was absolutely free of them which indicates a full rejection of these compounds by the membrane. As expected, the AMBR bulk and permeate also did not contain any of these. These molecules eluting before 9 minutes are supposedly huge, but their MW cannot be determined accurately. Nevertheless an extrapolation of the calibration curve would show that organics eluting at 8.4 minutes would have a MW of 10,200 kDa, but this should be interpreted with caution since they are far out of the calibration range (the standard of 219.3 kDa eluted at 12.3 minutes).

The medium MW compounds (11 to 13.2 minutes elution time) were in the range ≈800 to ≈90 kDa. It can be seen from Figure 29, Panel A that the absorbance of these compounds is greater in the SAMBRs than in the HR as these medium MW compounds are likely to be the result of the hydrolysis of the aforementioned high MW compounds eluting before 9 minutes. Thus the medium MW compounds appear more concentrated than in the leachate due to a further hydrolysis and stabilization of the leachate in the SAMBRs. The stabilization of recalcitrant organics can also result in a higher degree of aromaticity that would also translate to a higher absorbance at 254 nm.

Interestingly, the absorbance in the SAMBRs permeate is lower which is evidence that these medium MW compounds were rejected by the membrane to some extent. For instance on day 192, based on the relative absorbance, there was a 16 and 18% rejection in SAMBR1 and SAMBR2, respectively. Furthermore, the permeate peaks are sharper than the bulk peaks and the HR effluent peak; it can be seen that SAMBR2 bulk has a wide peak that plateaus between 11.2 and 12.2 minutes while its permeate counterpart is between 11.7 and 12.2, thus showing the rejection of the highest MW compounds amongst the medium MW compounds present in the bulk. The highest standard (219.3 kDa) eluted at 12.3, but nonetheless if the calibration curve is extrapolated beyond 12.3 minutes, it is calculated that elution times of 11.2 and 11.7 would correspond to a MW of 646 and 395 kDa, respectively. As a result, the compounds in that range of 395 - 646 kDa were more likely to be rejected by the membrane while the MW ≤ 395 kDa were observed in the permeate.

The comparison between the "SAMBR2 out" and "AMBR bulk" curves indicates firstly that medium MW compounds are further hydrolysed in the AMBR because they appear at a lower absorbance, therefore at a lower concentration than in SAMBR2 permeate. Secondly, there is a slight shift of the AMBR curves to the right (t = 12.7 min) compared to the SAMBR2 permeate curve that peak at ≈ 11.8 - 12 min meaning that the medium MW molecules are further broken down in the AMBR.

In terms of MW, it can be stated that medium MW compounds in SAMBR2 permeate are in the range 358 - 294 ($11.8 \leq t \leq 12$ min), while the AMBR bulk contained compounds of MW in the range 198-134 kDa ($12.4 \leq t \leq 12.8$ min). The COD polishing in the AMBR was therefore efficient on compounds of MW in the range 200 - 350 kDa.

Finally, all the samples exhibited a sharp peak with an elution time between 17.1 and 17.8 minutes, which is between the elution times of the standard 1.9 kDa (t=17 min) and 0.194 kDa (19.2 min), and thus correspond to low MW compounds or fulvic acids. The difference between these standards is very small in terms of MW and it is assumed that unknown peaks between these two standards are not significantly different.

The low MW fraction of the HR effluent has a greater absorbance than in SAMBR1 and SAMBR2 bulks. This could be due to the higher concentration of aromatics being absorbed at 254 nm in the leachate compared to the SAMBRs bulks. The low MW fraction in the AMBR has a greater absorbance than in the SAMBR2 permeate which could be due to the higher aromaticity of the former as fulvic acids are further degraded in the AMBR. In the case of SAMBR2 and AMBR the absorbance was higher in the permeate than in the bulk which could mean that there was an increase in the concentration of these fulvic acids in the permeate. However, it is more likely that the diffusion through the biofilm enables them to be degraded further than in the bulk which provided a greater degree of aromaticity that translated to a higher absorbance at 254 nm and was therefore not due to an increase in concentration. It is noteworthy that this rejection of medium MW and the higher degree of aromaticity of low MW in the permeate of AMBR happened not only on day 192 but also on all the other days sampled (days 136, 150, 164, 178 and 199). Regarding the evolution of the medium and low MW compounds over the 200 days, it can be stated that overall there was no build up in the absorbance in any effluent (HR, SAMBR1, SAMBR2 and AMBR). The panel A of Figure 30 shows the evolution of the medium MW compounds, while panel B shows the low MW compounds.

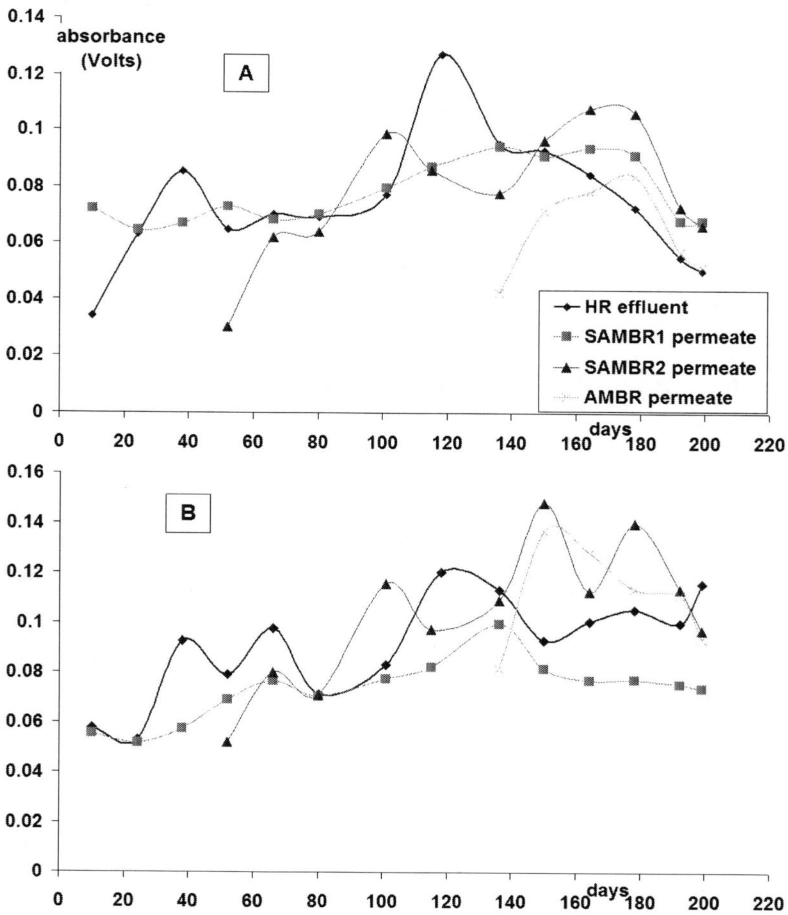

Figure 30. (A) Evolution with time of the absorbance of medium MW compounds (11 to 13.2 elution time, i.e., MW in the range 800 to 90 kDa) in the HR effluent and in the permeates of SAMBR1, SAMBR2 and AMBR. (B) Evolution with time of the absorbance of low MW compounds (17.1 to 17.8 elution time, i.e., MW in the range 1900 to 900 Da) in the HR effluent and in the permeates of SAMBR1, SAMBR2 and AMBR.

The medium MW compounds in SAMBR1 followed a very similar evolution compared to the SCOD; the absorbance increased slowly until day 136, then plateaued, and eventually, like the SCOD, there was a decrease in absorbance after day 178 probably due to a lower concentration of all organics caused by dilution with more fresh water at high OLRs. The evolution of medium MW compounds in the HR, SAMBR2 and AMBR was more chaotic but eventually their absorbance also decreased. Interestingly, the absorbance of SAMBR2 was two times lower than that of SAMBR1 during the start-up. This suggests that in SAMBR1 the competent population of hydrolytic bacteria acclimatised to the leachate medium was able to stabilize the organics by increasing their degree of aromaticity, while this was not the case in SAMBR2. At start-up SAMBR2 was virtually free of recalcitrant organics due to its inoculum and because the volume was adjusted with synthetic biomedium (Owen et al., 1979), and hence the concentration of organics with high aromaticity was lower. Similarly, the absorbance of the AMBR at start-up was considerably lower than in the other reactors because of the unadapted bacterial population that could not stabilize the refractory organics. In panel B for low MW compounds, the absorbance of SAMBR2 and AMBR was lower than that of SAMBR1, again due to the unadapted inoculum and the organics-free biomedium, but the difference was more significant for the medium MW compounds suggesting that the increase of SCOD in SAMBR2 during start-up was mainly due to the build-up of medium MW compounds that could not be hydrolyzed, rather than the fulvic acids.

3.7. Gas Chromatography - Mass Spectrometry

3.7.1. Introduction

This technique can provide us with novel insights into the processes taking place in the continuous treatment of OFMSW in two-stage membrane reactors. However, it should be born in mind that library identification should be interpreted with caution as sometimes a high match can be wrong, and this can be detected by comparing the mass spectra of the peak selected with the mass spectra of the proposed compounds. Moreover, a calibration line was drawn for o-hydroxybiphenyl and Bis (2-ethylhexyl)phthalate in order to gain more information on the concentrations of these specific recalcitrants found in the reactors. The raw chromatographs can be seen in Figure 31. A control sample (referred as 'scrap') consisted of 500 mL of DW in which small pieces of the plastic (plastic scrap) used to make the reactor were added and the mixture was shaken for few weeks at 30°C in order to determine which components if any could leach from the reactor's construction material (Table 11). Table 12 to Table 15 gather the peak identification number, the match percentage, the retention time, the area and the name of the components that were detected in the effluent of each reactor, but not in the blank (DW that followed the same SPE protocol) nor the sample with plastic scraps. In addition, the last column comments on the biodegradability by comparing the areas of the respective peaks.

Table 11. Compounds detected by GC-MS in the reactor's plastic scraps

Percentage match	Retention time (min)	Compound	Formula
78	11.808	diphenyl ether	$C_{12}H_{10}O$
83	13	dimethylphthalate	$C_{10}H_{10}O_4$
97	14.5	hexadecane	$C_{16}H_{34}$
93	15.304	diethylphthalate	$C_{12}H_{14}O_4$
97	16	Carbamic acid, butyl-,3-iodo-2-propynyl	$C_8H_{12}INO_2$
83	18.901	Diisobutyl phthalate	$C_{16}H_{22}O_4$
99	19.2	hexadecanoic acid, methyl ester	$C_{17}H_{34}O_2$
91	20.107	dibutylphthalate	$C_{16}H_{22}O_4$
92	21.198	9-octadecenoic acid, methyl ester (E)	$C_{19}H_{36}O_2$
78	26.5	Dicyclohexyl phthalate	$C_{20}H_{26}O_4$
91	28.5	decanedioic acid, bis (2-ethylhexyl)ester	$C_{26}H_{50}O_4$
98	28.725	13-Docosenamide, (Z)	$C_{22}H_{43}NO$

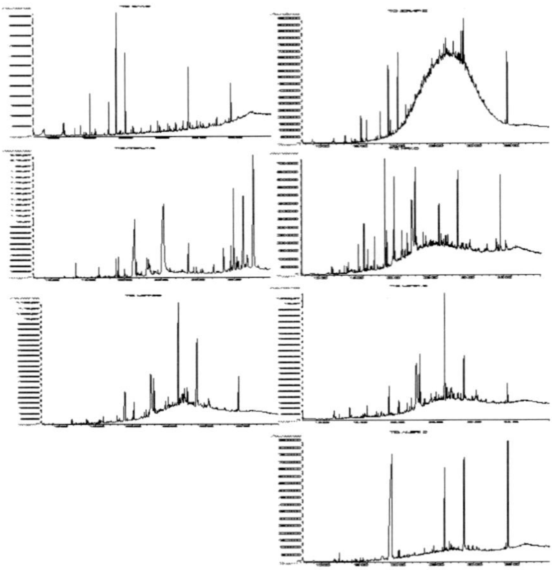

Figure 31. GC-MS chromatographs. From left to right: blank, reactor's plastic scrap, deionized water with anti-foaming agent, effluent of the hydrolytic reactor, SAMBR1 permeate, SAMBR2 permeate, AMBR permeate.

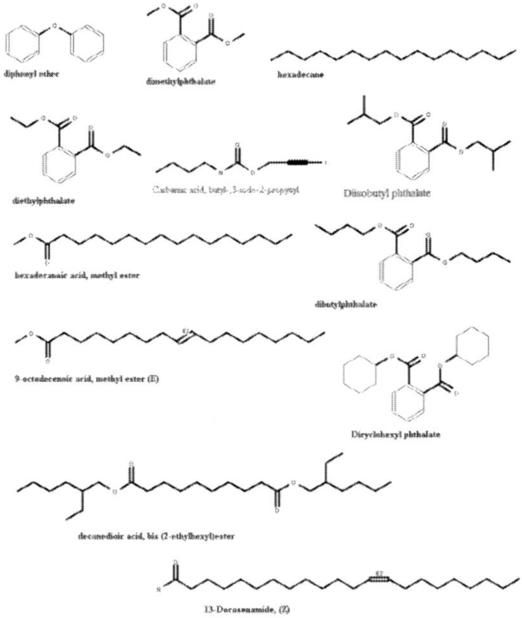

Figure 32. Compounds detected by GC-MS in the reactor's plastic scraps.

Table 12. Recalcitrant compounds detected by GC-MS in the effluent from the HR. In the case of unidentified compounds, the retention index of the peak is given. Reprinted from (Trzcinski and Stuckey, 2009b)

Peak ID	match %	Retention time (min)	Recalcitrants in HR effluent	Formula	Peak Area
1	93	13.357	butylated hydroxytoluene	$C_{15}H_{24}O$	12,281,748
2	94	13.72	o-hydroxybiphenyl	$C_{12}H_{10}O$	16,062,321
3		13.91	1462.7		26,872,283
4	93	16.033	benzophenore	$C_{13}H_{10}O$	112,030,035
5	93	16.525	tridecanoic acid, 12-methyl, methyl ester	$C_{15}H_{30}O_2$	34,198,411
6		17.486	1691.5		56,558,115
7	99	19.141	pentadecanoic acid, 14-methyl, methyl ester	$C_{17}H_{34}O_2$	62,748,504
8	99	21.113	8,11-octadecadienoic acid, methyl ester	$C_{19}H_{34}O_2$	38,295,039
9		21.271	1964.1		46,078,046
10		21.852	2008.9		95,664,113
11	94	22.481	phenol 4,40-(1-methylethylidene)bis	$C_{15}H_{16}O_2$	570,361,325
12		22.978	2096.1		411,285,760
13	87	24.052	padimate O	$C_{17}H_{27}NO_2$	101,864,975
14	95	24.39	1-phenanthrene carboxylic acid 1,2,3 m,4,4a,9,10,10a-octahydro-1, 4a-dimethyl-7-(1-methylethyl)-,methyl ester,[1R(lalpha,4abeta, 10aalpha)]	$C_{21}H_{27}O_3$	93,127,843
15	72	26.179	Bis (2-ethylhexyl)phthalate	$C_{24}H_{38}O_4$	171,161,815

3.7.2. HR Effluent

The analysis revealed that most of the compounds were below 10 μg/L while only two compounds (Bisphenol A and unidentified compounds with RI = 2096) were in the range 14 - 38 μg/L. The chromatograms revealed that butylated hydroxytoluene found in the HR effluent was completely degradable because they were not found in both SAMBRs and the AMBR effluents. However, previous work has shown that butylated hydroxytoluene can leach from plastic and tubings (Shpiner, 2007), but it was not detected in our control sample containing plastic scraps. Similarly, 8,11-octa decadienoic acid, methyl ester (peak ID 8), pentadecanoic acid, 14-methyl, methyl ester (peak ID 7) and tridecanoic acid, 12-methyl-, methyl ester (peak ID 5) were three aliphatic molecules that were not detected in the SAMBR permeates due to their complete degradation in this reactor. Surprisingly, padimate O and the phenanthrene carboxylic acid that are aromatics and thus considered as difficult to biodegrade were successfully degraded in the SAMBRs due to the complete retention of bacteria. Most of PAHs (polycyclic aromatic hydrocarbons) in leachate are stable and it is difficult to cleave the ring without oxygen. However, recent research has shown that unsubstituted low molecular polycyclic aromatic compounds can be degraded under nitrate-reducing, iron-reducing, sulfate-reducing and methanogenic conditions (MacRae and Hall, 1998; Xu et al., 2008). MacRae and Hall (1998) showed that PAHs could be degraded under denitrifying conditions as long as other nutrients were not limiting. The half-lives of low molecular weight PAH ranged from approximately 33-88 days. Degradation of high molecular weight PAH was slower, or not observed. Half-lives ranged from 143-812 days. Thus, the presence of nitrate from the permeate of the AMBR may have assisted the degradation of the padimate and the phenanthrene under nitrate-reducing conditions in the HR.

3.7.3. SAMBR1 and 2 Permeates

Table 12 to Table 14 show that o-hydroxybiphenyl and Bisphenol A can be considered as non-biodegradable anaerobically because their areas increased by 9 and 12% in SAMBR1, respectively. In SAMBR2 they increased by 123 and 15%, respectively. The concentration of o-hydroxybiphenyl was found to be 10.2, 11.2 and 22.8 μg/L in the HR, SAMBR1 and SAMBR2, respectively. Bisphenol A (CAS 80057) is a very common compound found in landfill leachate and it is known as a potentially endocrine disrupting industrial chemical (Schwarzbauer et al.,

2002; Wintgens et al., 2004). It is found in epoxy resin and is used as a stabilizer for polyvinyl chloride resin. Yasuhara et al. (1999) found that among more than 100 organic compounds, Bisphenol A was at a very high concentration of 61.4 mg/L. Wintgens et al. (2004) and Yamamoto et al. (2001) also found concentrations ranging up to milligrams per litre in raw landfill leachate. Wintgens et al. (2004) stated that the high concentrations can be explained by leaching effects from polycarbonate plastics abundant in the deposited waste. The conditions encountered in the landfill in terms of alkaline pH, mesophilic temperature and high ionic strength might promote the leaching process. They used a pilot plant MBR configuration including denitrification and two-stage nitrification with side-stream tubular ultrafiltration that rejected 99% of bisphenol A. They reckon that it seems to be removed most effectively by biodegradation in a system with high sludge age and complete retention of solids.

Several authors have observed that the concentration of Bisphenol A tends to decrease over time in a landfill (Asakura et al., 2004). In our study, bisphenol A was the main peak in the HR effluent at a concentration estimated to be in the range 19-38 μg/L, while in the SAMBR1 and SAMBR2 its concentration increased to 21-44 μg/L. On the other hand, these authors have observed that the concentration of Bis (2-ethylhexyl)phthalate remained at the same level. In our study it was not detected in the blank and scrap, and its area more than doubled from the HR effluent to SAMBR1 (+283%) and SAMBR2 (+179%) permeates, suggesting that it could be secreted by bacteria themselves, or is the catabolic end product of non-detected compounds. A calibration line was obtained for Bis (2-ethylhexyl)phthalate ($R^2 = 0.99$) and it revealed that the concentration in SAMBR 1 and SAMBR2 was 1.4 and 1 mg/L, respectively. The concentration in the HR and the AMBR was found to be 0.33 and 0.3 mg/l, respectively. Some molecules were found to be slowly biodegradable because their areas decreased when passing through both SAMBRs. These molecules were benzophenone which decreased by 81 and 73% in SAMBR1 and 2, respectively, plus unidentified compounds with retention indices 1462.7 and 2096.1 in SAMBR1 and 2008.9 and 2096.1 in SAMBR2.

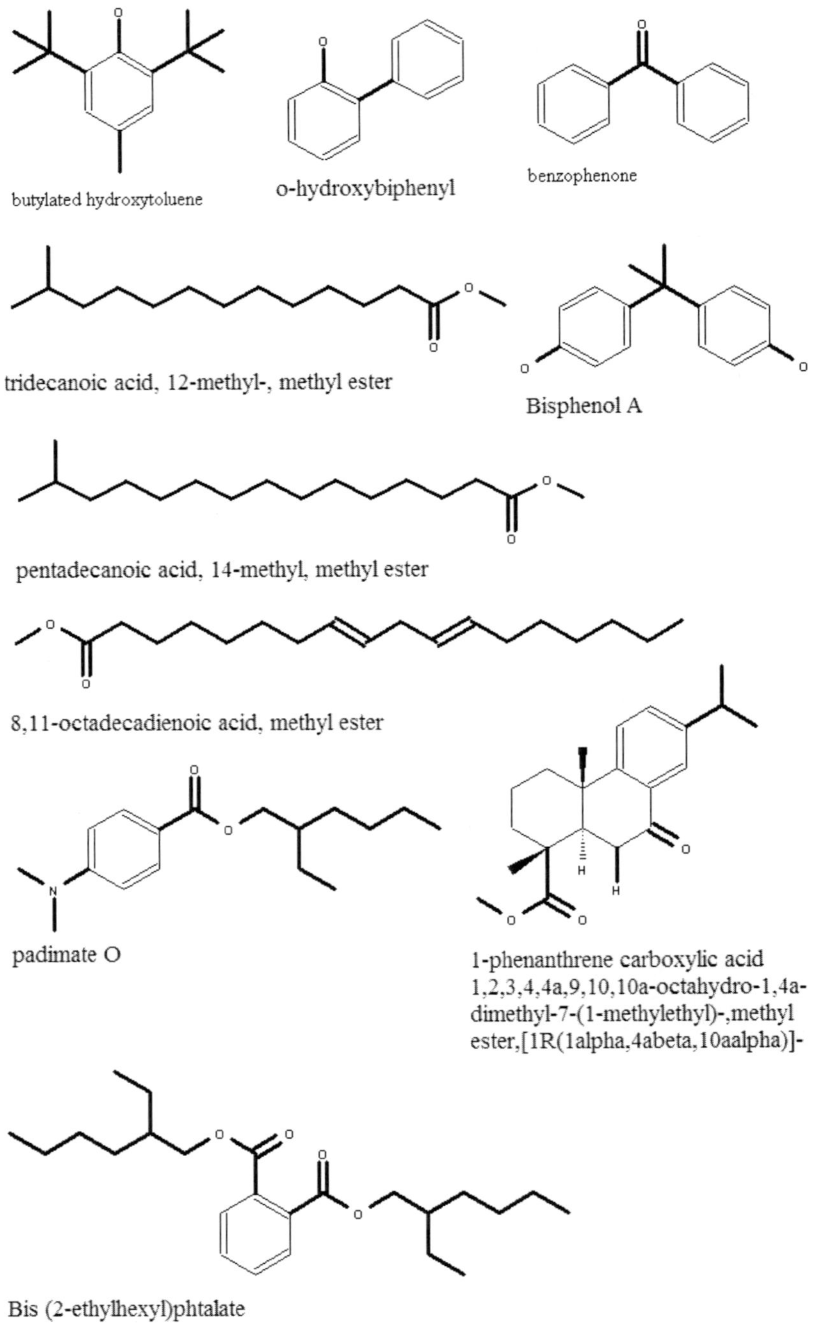

Figure 33. Compounds detected by GC-MS in the HR effluent.

3.7.4. SAMBR2 and AMBR Permeates

In comparing the SAMBR2 and AMBR permeates, it can be seen that phenol 2,4-bis(1,1-dimethylethyl) and N-phenethylbenzenesulfonamide were not degraded aerobically because their areas were found to increase when passing through the AMBR: their areas increased by 45 and 786%, respectively. The concentration of the latter reached 44-87 μg/L in the AMBR. Mansouri et al. (2007) also found that phenol 2,4-bis(1,1-dimethylethyl) was among the components that remained after an aerobic fixed bed process treating landfill leachate. As Bisphenol A was not found in the AMBR permeate, it can be stated that it was fully biodegraded aerobically. This is in line with Asakura et al. (2004) who treated raw leachate by aeration and found that bisphenol A decreased from 70.9 - 224 μg/l to 0.11- 0.24 μg/l in the effluent. This is also in line with the Biocatalysis / Biodegradation database from the University of Minnesota (http://umbbd.msi.umn.edu/index.html) that states that Bisphenol A is metabolized by a Gram-negative aerobic bacterium (strain MV1) via a novel pathway. The metabolism involves two primary pathways: a minor pathway, in which Bisphenol A is first oxidised to a triol, and a major pathway, in which Bisphenol A is metabolized to form the intermediates 4-hydroxybenzoate and 4-hydroxyacetophenone that are converted to carbon dioxide and biomass (Spivack et al., 1994). The former decreased by 68% whereas the latter decreased by 67% in the AMBR. Bis (2-ethylhexyl)phthalate often referred as DEHP is the most common plasticizer (Bauer and Herrmann, 1997).

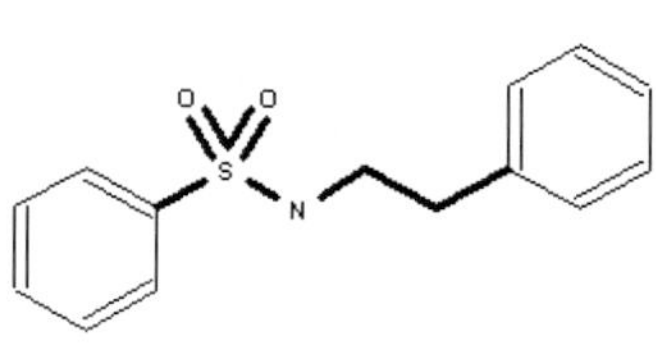

N-phenethylbenzenesulfonamide

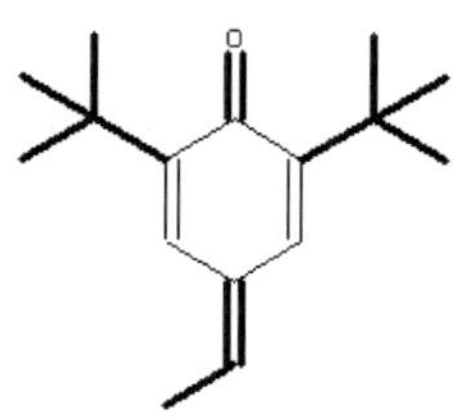

2,5-cyclohexadien-1-one,2,6 bis(1,1-dimethylethyl)-4-ethylidene-

Figure 34. New compounds detected by GC-MS in SAMBR1 permeate.

Table 13. Recalcitrant compounds detected by GC-MS in the permeate of SAMBR1. In the case of unidentified compounds, the retention index of the peak is given. The last column comments on the anaerobic biodegradability by comparing the peak area of the same compound in the HR effluent. NB = Non-biodegradable, B = Biodegradable

Peak ID	match %	Ret. time	Recalcitrants in SAMBR1 permeate	Formula	Peak Area	Anaerobic Biodegra-dability
16		11.68	1333.7		26,996,170	New
2	94	13.72	o-hydroxybiphenyl	$C_{12}H_{10}O$	17,502,568	NB (+9%)
3		13.91	1462.7		14,119,684	B (-48%)
4	83	16.03	benzophenone	$C_{13}H_{10}O$	21,735,256	B (-81%)
17		18.21	1742.3	$C_{15}H_{3}O_{2}$	15,836,896	New
18	78	18.99	n-phenethyl-benznesulfonamide		208,608,266	New
19	78	20.23	2,5-cyclohexadien-1-one,2,6 bis(1,1-dimethylethyl)-4-ethylidene-	$C_{14}H_{15}NO_{2}S$	44,712,759	New
11	94	22.48	Phenol 4,40-(1-methylethyli-dene)bis	$C_{16}H_{24}O$	640,395,425	NB (+12%)
12		22.98	2096.1		179,375,192	B (-56%)
15	91	26.18	Bis (2-ethylhexyl)-phthalate	$C_{24}H_{38}O_{4}$	655,807,845	NB (+283%)
20	95	27.133	Phosphine imide, P,P,P-triphenyl	$C_{18}H_{16}NP$	123,496,720	New

Other intermediates of the degradation pathway of Bisphenol A include toluene and vanillin. Interestingly, some molecules were found to be non-biodegradable in an anaerobic environment, but could be slowly biodegraded in the AMBR such as 2,5-cyclohexadien-1-one,2,6 bis(1,1-dimethylethyl)-4-ethylidene- and Bis (2-ethylhexyl)phthalate.

Table 14. Recalcitrant compounds detected by GC-MS in the permeate of SAMBR2. In the case of unidentified compounds, the retention index of the peak is given. The last column comments on the anaerobic biodegradability by comparing the peak area of the same compound in the HR effluent. NB = Non-biodegradable, B = Biodegradable

Peak ID	match %	Ret. time	Recalcitrants in SAMBR2 permeate	Formula	Peak Area	Anaerobic Biodegra-dability
21	93	12.4	phenol 2,4-bis(1,1-dimethylethyl)	$C_{14}H_{22}O$	11,305,194	New
2	94	13.72	o-hydroxybiphenyl	$C_{12}H_{10}O$	35,846,178	NB (+123%)
22	94	15.69	benzenemethanol, alpha-phenyl	$C_{13}H_{12}O$	36,535,074	New
4	94	16.03	benzophenone	$C_{13}H_{10}O$	29,918,768	B (-73%)
18	78	18.99	N-phenethyl-benzenesulfonamide	$C_{14}H_{15}NO_2S$	147,543,256	New
19	78	20.23	2,5-cyclohexadien-1-one,2,6bis(1,1-dimethylethyl)-4-ethylidene-	$C_{16}H_{24}O$	44,712,759	New
10		21.85	2008.9		46,417,418	B (-50%)
11	95	22.48	phenol 4,40-(1-methylethylidene)bis	$C_{15}H_{17}O_2$	655,737,185	NB (+15%)
12		22.98	2096.1		262,561,636	B (-36%)
15	91	26.18	Bis (2-ethylhexyl)phthalate	$C_{24}H_{38}O_4$	478,178,027	NB (+179%)

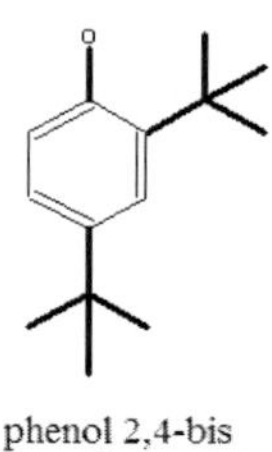

phenol 2,4-bis (1,1-dimethylethyl)

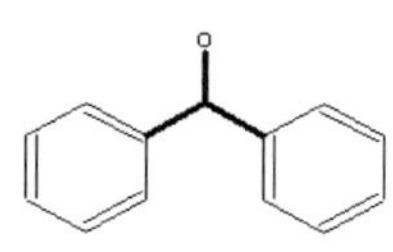

benzenemethanol, alpha-phenyl

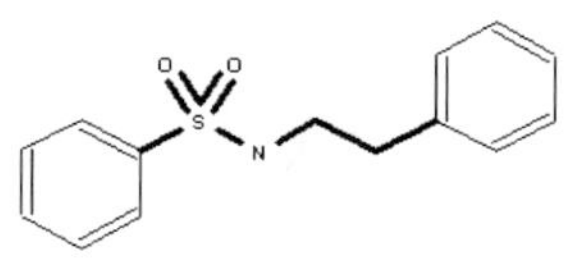

N-phenethylbenzenesulfonamide

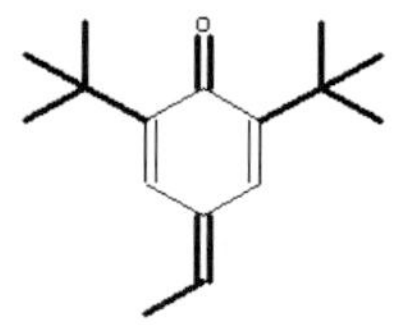

2,5-cyclohexadien-1-one,2,6 bis(1,1-dimethylethyl)-4-ethylidene-

Figure 35. New compounds detected by GC-MS in the permeate of SAMBR2.

Several authors have observed that the concentration of Bis (2-ethylhexyl)phthalate remains constant in a landfill (Asakura et al., 2004; Sakamoto et al., 2000). Because Bis (2-ethylhexyl)phthalate has a high octanol/water partition coefficient, it can be considered as hydrophobic and therefore is expected to be removed when suspended solids are removed. In other words it tends to attach to biomass more than hydrophilic compounds. Asakura et al. (2004) found that it could not be removed by aeration, coagulation/sedimentation nor biological treatment. It has been reported elsewhere that DEHP can be degraded by a common soil bacterium (*Rhodococcus rhodochrous*) with an easily used carbon source, and that the final metabolites are 2-ethylhexanol, 2-ethylhexanoic acid and phthalic acid (Horn et al., 2004; Nalli et al., 2003). Moreover, 2-ethylhexanoic acid was observed to be particularly resistant to further degradation and exhibited acute aquatic toxicity. However, in our study the DEHP concentration increased in the SAMBRs, whereas its concentration was lower in the AMBR, but none of the metabolites given in the paper of Horn et al. (2004) were found. This could be due to a different metabolic pathway in the AMBR or because of their hydrophobicity; these metabolites were preferentially attached to the sludge and hence were not detected. Nevertheless, new molecules appeared in the AMBR permeate such as diisobutyl phthalate and an unidentified peak with a retention index of 1510.6. Moreover, the calibration line of Bis (2-ethylhexyl)phthalate was used to calculate the concentration of recalcitrants in the SAMBR2 and AMBR permeate. Then the COD equivalence of each compounds were found to be very low when added together. As a matter of fact, the COD concentrations of recalcitrants in SAMBR2 and AMBR permeate was found to be equal to 17.7 and 20 mg/L, respectively, which contrast with the respective total COD concentrations of about 400 and 300 mg/L. Section 3.4.1 contains details indicating that recalcitrant proteinaceous compounds in the SAMBR2 and AMBR permeates accounted for about 100 and 50 mg/l, which is equal to 150 and 75 mg COD/L. Therefore, about 250 and 225 mg/L of the permeates are still unknown compounds. The use of GC-MS is limited to the identification of non-polar, volatile and thermostable compounds which explains why so little information was found in this study. Techniques such as LC-MS or Matrix Assisted Laser Desorption Ionization-Time of Flight-Mass Spectrometry (MALDI-ToF-MS) would certainly shed more light on the nature of the high MW compounds that were not detected.

Table 15. Recalcitrant compounds detected by GC-MS in the permeate of the AMBR. In the case of unidentified compounds, the retention index of the peak is given. The last column comments on the aerobic biodegradability by comparing the peak area of the same compound in the SAMBR2 permeate. NB = Non-biodegradable, B = Biodegradable. Reprinted from (Trzcinski and Stuckey, 2009b)

Peak ID	match %	Ret. time	Recalcitrants in AMBR permeate	Formula	Peak Area	Anaerobic Biodegra-dability
21	91	12.4	phenol 2,4-bis (1,1-dimethylethyl)	$C_{14}H_{22}O$	16,343,590	NB (+45%)
23		14.7	1510.7			New
18	78	18.99	N-phenethyl-benzenesulfonamide	$C_{14}H_{15}NO_2S$	1,307,129,372	NB (+786%)
24	81	20.12	1,2-benzenedi carboxylic acid, butyl (2-methylpropyl)ester	$C_{16}H_{22}O_4$	21,287,952	New
19	78	20.23	2,5-cyclohexadien-1-one, 2,6bis (1,1-dimethylethyl)-4-ethylidene-	$C_{16}H_{24}O$	14,309,919	B (-68%)
15	91	26.18	Bis (2-ethylhexyl)phthalate	$C_{16}H_{38}O_4$	156,613,458	B (-67%)

3.7.5. Phthalates and Plasticizers

Plasticizers are compounds that are added to polymers in order to improve the properties of a plastic such as increasing its flexibility, and several phthalates were detected in this study. It has been demonstrated that plasticizers tend to leach from solid polymer matrices into the environment (Fromme et al., 2002). For instance dimethylphthalate was found in the reactor plastic scrap but was not detected in the reactor indicating that it could be readily biodegraded. Diethylphthalate was found at a similar level in the scrap and in the HR effluent. The fact that it was not detected in the SAMBR permeates indicates that it could be biodegraded completely thanks to the long solid retention times achieved in SAMBRs. Dibutylphtalate was found in the anaerobic reactors but also in the scrap suggesting that it might come from the reactors plastic. Interestingly, its area decreased greatly in the SAMBRs (from 270 million in HR effluent to 25.6 and 19.9 million in SAMBR1 and 2 permeate, respectively), and was absent in the AMBR indicating that a great proportion of it can be degraded anaerobically and totally degraded aerobically.

3.8. Denaturing Gel Gradient Electrophoresis (DGGE)

3.8.1. Archaea

Archaeal DNA was amplified using the nested polymerase chain reaction (PCR) reaction. First, the DNA template was amplified using the primers 46F and 1017R. Then 1 microliter PCR product was used for the second PCR reaction using the primers 344FGC and Univ522. The objective was to investigate the differences if any between the archaeal population in the HR and the SAMBR and the effect of the HRT on their evolution. As Figure 36 shows, there were only three visible bands and with the exception of lane 5 the bands were more dominant in SAMBR1 (Lanes 5 to 10). Previous trials with archaeal primers gave very faint bands as is in Figure 36. Therefore, to be able to cut out some bands and identify any species, it was decided to run three PCR reactions in parallel for the sample SAMBR1 day 137 (lane 8) and to load the 3 times concentrated PCR product on the DGGE gel and thereby obtaining brighter bands. This was indeed a good but time-consuming method to get bright bands as it can be seen that bands in lane 8 (SAMBR1, day 137) are much brighter than in the other lanes. It is believed that the bands in other lanes were very faint because of the small amount of DNA template used in the PCR reaction; as

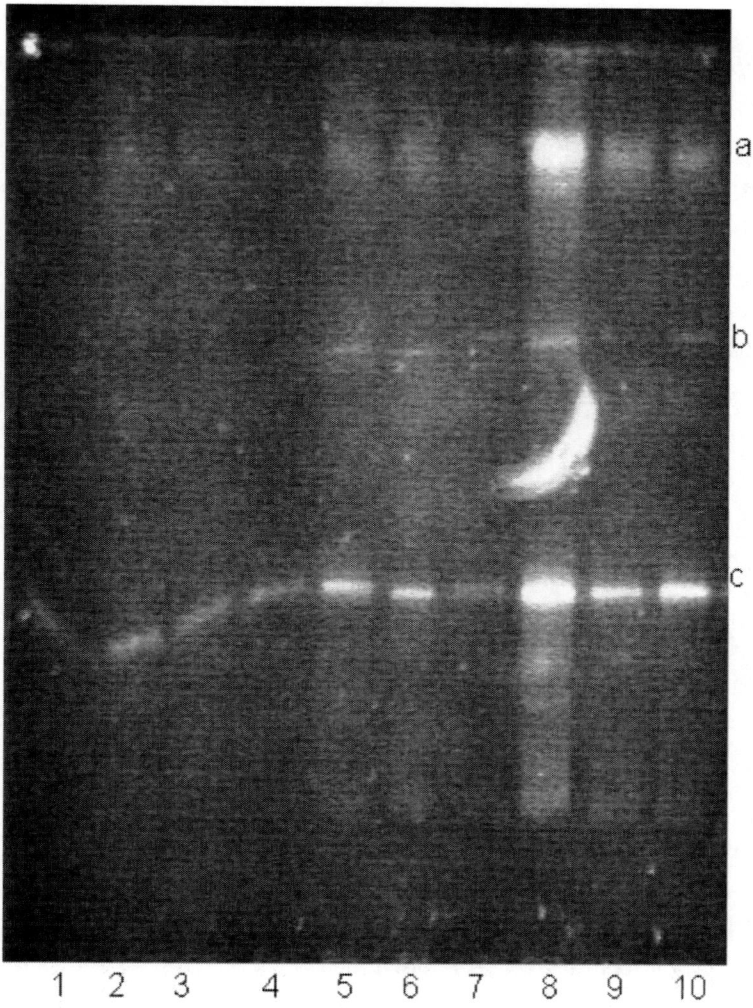

Figure 36. Evolution with time of the archaeal population in the HR (Lane 1 to 5) and SAMBR1 (Lane 6 to 10). Lanes 1 and 6 = day 76; Lanes 2 and 7 = day 119; Lane 3 and 8 = day 137; Lane 4 and 9 = day 178; Lane 5 and 10 = day 203.

a result all the archaeal species present in the bioreactor did not appear on the DGGE gel. Nevertheless, the presence of bands indicates the predominant species in the reactor. Bands a,b and c were cut from lane 8 but only band b could be identified successfully as an organism from the genus *Methanothrix* named *Methanosaeta sp.* clone DI C03 (Accession number AY454761.1, 100% match, E value = 0.82). Because *Methanosarcina* was also found in the experiment described in the next chapters, it was hypothesized that band c corresponded to *Methanosarcina*. If this is true it would mean that *Methanosarcina sp.* develop preferentially in the HR where the acetate concentration is higher, whereas *Methanosaeta* would be dominant in the SAMBR where acetate concentration is very low. This is in line with Ehlinger et al. (1987) who stated that *Methanosarcina sp.* develop preferentially throughout or in the inlet of reactors, wherever the acetate concentration was equal to or higher than 350 mg L^{-1}. When the acetate concentration falls below 350 mg/L, *Methanothrix sp.* becomes the prevailing species. In the HR (lanes 1 to 5), only band c could be seen except on day 203 (Lane 5) where bands a and b also appear. The appearance of *Methanosaeta* in the HR could be due to the low acetate concentration towards the end (below).

3.8.2. Bacteria

Bacterial DNA was amplified using a nested PCR reaction. First, the DNA was amplified using the primers 27F and 1492R. Then 1 microliter of PCR product was used as template for the second PCR using the primers 338FGC and 518R. Again the bands were quite faint even with triplicate PCR products with this combination of primers. Nevertheless, it can be seen from Figure 37 that the SAMBR2 started up on day 46 (Lane 5) with no dominant bacteria although that sample contained DNA as it was checked on agarose gel beforehand. On day 119 (Lane 4), three distinct dominant species of bacteria (bands a, c and d) could be distinguished, and as the HRT was reduced the number of bands also diminished so that only two bands (c and d) were visible on day 137 (lane 3) and later only 1 band (b) in lanes 1 and 2. Moreover, there was a significant shift in bacterial population as the HRT was decreased because band b appeared only in lane 1 and 2, i.e., when the HRT was less than 4 days. Bands a, b, c and d were cut and after cloning and sequencing as described in Materials and Methods, Table 16 was drawn. As the bands on DGGE gel were quite faint using the nested PCR method mentioned above, another combination of primers was tested: 341FGC and 907R.

Table 16. Bands cut from the DGGE gel using bacterial primers

Band	Closest Match	Accession Number	Genus	Match %	E value
a	Uncultured bacterium	U81676.2	Uncultured Bacteroidales	100	1.10^{-95}
b	Uncultured bacterium	EF589977.2	Uncultured Bacteroidales	98	2.10^{-92}
c	Uncultured bacterium	AY693829.1	Uncultured Bacteroidales	100	1.10^{-95}
d	Uncultured bacterium	EF559220.1	Uncultured Bacteroidales	100	1.10^{-95}

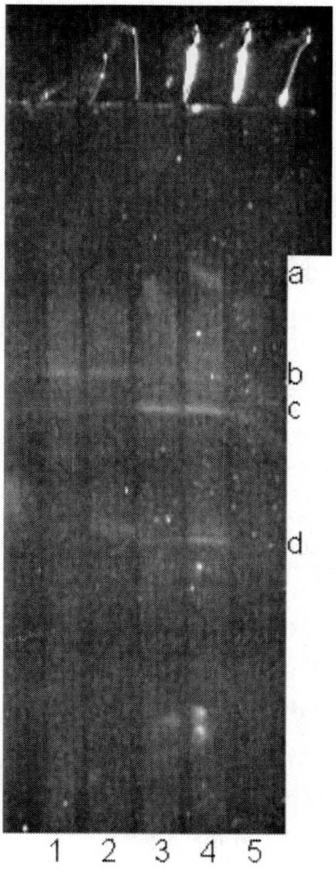

Figure 37. Evolution with time of the bacterial population in SAMBR2 using nested PCR 27F and 1492R then 338FGC and 518R. Lane 1 = day 203; Lane 2 = day 178; Lane 3 = day 137; Lane 4 = day 119; Lane 5 = day 46.

Figure 38 shows the same samples amplified with the nested PCR in triplicate on the left hand side and the samples amplified using 341FGC and 907R on the right hand side. It is clear that the latter combination gave rise to much brighter bands on an agarose gel, and the PCR products were therefore loaded on a DGGE gel in Figure 38. It can be seen from Figure 39 that there is no band in lane 11 (day 46) which is when SAMBR2 was started.

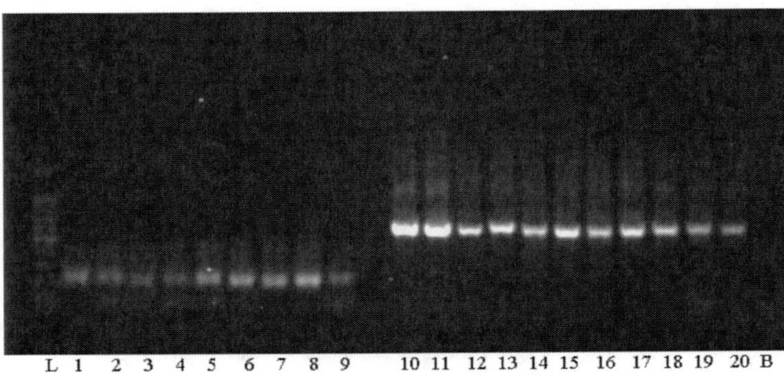

Figure 38. Comparison between the nested PCR products (27F and 1492R then 338FGC and 518R) on the left hand side and the PCR products using primers 341FGC + 907R on the right hand side. L=Ladder; Lane 1=SAMBR1 day119; Lane 2=SAMBR1 day 137; Lane 3=SAMBR1 day 178; Lane 4=SAMBR1 day 203; Lane 5=HR day 76; Lane 6=HR day 119; Lane 7=HR day 137; Lane 8=HR day 178; Lane 9=HR day 203; Lane 10=HR day 137; Lane 11=HR day 178; Lane 12=HR day 203; Lane 13=SAMBR1 day 137; Lane 14=SAMBR1 day 178; Lane 15=SAMBR2 day 137; Lane 16=SAMBR2 day 178; Lane 17=SAMBR2 day 203; Lane 18=AMBR day 137; Lane 19=AMBR day 178; Lane 20=AMBR day 203; B=blank.

The bands could be seen only from lane 12 (day76) onwards and became brighter on lane 13 (day119). This highlights the fact that SAMBR2 did not start with the proper bacterial population and it took at least 30 days to get a similar population as SAMBR1 in lane 6. This lack of bacteria for start-up was correlated with an increase in bulk SCOD as explained above. Unfortunately, the bright band at the top of the gel which is common to all lanes indicates that some DNA could not migrate through the gel. This is because 40% as a lower gradient was already too high.

Another gel with a denaturing gradient of 30-70% was run but the bands were all compressed at the top of the gel. An optimum gradient for these PCR products would have been 20-60%.

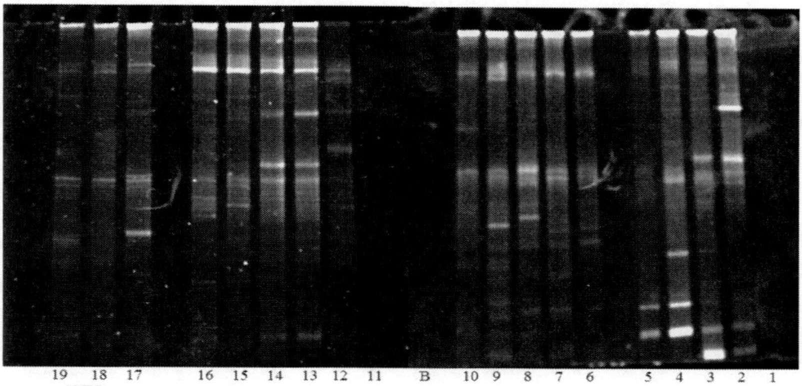

Figure 39. Evolution with time of the bacterial population in the HR (Lanes 1 to 5), SAMBR1 (Lanes 6 to 10), SAMBR2 (Lanes 11 to 16) and AMBR (Lanes 17 to 19) using primers 341FGC and 907R. Lane 1=HR day 76; Lane 2=HR day 119; Lane 3=HR day 137; Lane 4=HR day 178; Lane 5=HR day 203; Lane 6=SAMBR1 day 76; Lane 7=SAMBR1 day 119; Lane 8=SAMBR1 day 137; Lane 9=SAMBR1 day 178; Lane 10=SAMBR1 day 203; B=blank; Lane 11=SAMBR2 day 46; Lane 12=SAMBR2 day 76; Lane 13=SAMBR2 day 119; Lane 14=SAMBR2 day 137; Lane 15=SAMBR2 day 178; Lane 16=SAMBR2 day 203; Lane 17=AMBR day 137; Lane 18=AMBR day 178; Lane 19=AMBR day 203.

3.8.3. Ammonia - Oxidisers

The DNA from ammonia-oxidising bacteria (AOB) was amplified using the primers CTO189F-GC and CTO654R. Bands a, b, c and d were cut from the DGGE gel in Figure 40 and it was found that they all belong to the genus *Nitrosomonas*, except band d which belong to the genus *Nitrosospira*. The details of the bands are given in Table 17. It is clear from Figure 40 that there was no AOB species in Lane 4 which corresponds to a few days after the AMBR was started up. This explains why there was no ammonia conversion in the AMBR at the beginning. Then on lane 5 (day 178), two clear bands a (uncultured *Nitrosomonas sp.*) and b (*Nitrosomonas europaea* ATCC 19718) can be seen as well as a faint band c (*Nitrosomonas* sp. NM 41). Eventually on day 203 (lane 6), bands b and c became more dominant as the HRT was below 1 day. It is noteworthy that bands a, b and c are all from the genus *Nitrosomonas* and all are *Nitrosomonas europaea* - like species which are well known AOB. Interestingly, a band was found in the HR but only on day 203, at the end of the run. This band d was found to be *Nitrosospira sp.* Nsp40 which is an AOB and was not found in the AMBR. This is, however, controversial as heterotrophs are known to be fast growing microorganisms and AOB would have been unable to compete with them for oxygen present in the feedstock fed once a day in the HR. It is suggested that at low DO levels, AOB used nitrite or nitrate as an artificial electron acceptor and generate nitrous oxide gas (Ritchie and Nicholas, 1972).

Table 17. Bands cut from the DGGE gel using AOB primers

Band	Closest Match	Accession Number	Genus	Match %	E value
a	*Uncultured Nitrosomonas sp.*	AJ307985.1	*Nitrosomonas*	98	0
b	*Nitrosomonas europaea ATCC 19718*	AL954747.1	*Nitrosomonas*	98	0
c	*Nitrosomonas sp. NM41*	AF272421.1	*Nitrosomonas*	97	0
d	*Nitrosospira sp. Nsp40*	AY123787.1	*Nitrosospira*	100	0

The *Nitrospira* and *Nitrobacter* NOB species were not detected because the primers used were specific to AOB. However, these NOB species were expected to be present in the AMBR because nitrite conversion to nitrate was observed from day 167 onwards. To gain more insight into the nitrification process, it would have been useful to use primers specific to NOB such as NSR1113F and NSR1264R for *Nitrospira* (Wang et al., 2007). Alternatively, Wittebolle et al. (2008) have used a nested PCR where the first PCR amplifies specifically AOB whereas the second PCR is performed with total bacterial primers (338FGC and 518R) that will amplify NOB as well. It is possible that some AOB were missed due to the gradient (38-50%) used to make up the gel. Other studies have used larger gradients (45-60%) when using primers CTO189FGC and CTO654R (Wittebolle et al., 2008). It can be seen that band a is already at the top of the gel and it is therefore possible that the low gradient (38%) was too high and did not allow some bands to migrate and remained at the top of the gel. Also band d was at the bottom of the gel, and thus it is possible that some bands were missed because they migrate further than band d due to the high gradient (50%) being too low.

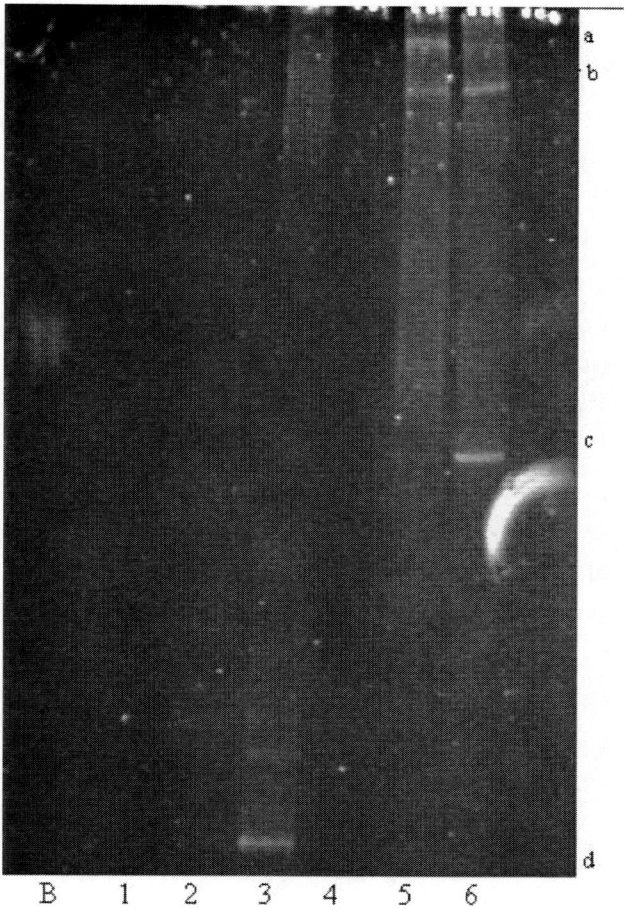

Figure 40. Evolution with time of the population of AOB in the HR and the AMBR using the primers CTO189FGC and 654R. B=Blank; Lane 1=HR day137; Lane 2=HR day178; Lane 3=HR day203; Lane 4=AMBR day137; Lane 5=AMBR day178; Lane 6=AMBR day203.

Conclusion

The following conclusions can be drawn from this study:

- The use of a membrane for leachate treatment had several advantages; the complete retention of bacteria allowed for stable operation as no VFAs accumulated even when propionate was the predominant acid in the leachate. The COD removal was greater than 90% at a HRT of 1.6-2.3 days at a maximum OLR of 20 g COD/L.day. Even though the influent COD of the leachate was constantly changing giving rise to a transitory F/M ratio over time, the permeate COD from the SAMBR was typically between 300 and 500 mg/L which can therefore be defined as a stabilised leachate. Because of the fluctuating properties of the leachate produced in the HR the process was deemed more representative because the SAMBR treated a leachate with varying organic strength which is what can happen on a full scale (Kraume et al., 2009).
- An inoculum fed on synthetic VFAs was not suitable for start-up because it did not contain hydrolytic and acidogenic bacteria for a leachate medium. As a result, a slow production of VFAs, and therefore a slow increase in methane production were observed. Furthermore, the bulk SCOD in SAMBR2 kept increasing during start-up, which strongly suggests that, for a lignocellulosic-based feed, it is paramount to start up the SAMBR with a hydrolytic population acclimatised to the leachate medium to avoid SCOD build up in the bulk. The DGGE picture confirmed this observation.
- The COD removal in SAMBR2 was 94.5% on average, and only 1.6% in the AMBR so that a total COD removal of 96.1% was achieved at 0.4 day HRT. On average 26% of the recalcitrants from SAMBR2 could be degraded aerobically in the AMBR.
- However, as in SAMBR1, at HRTs lower than 2 days, particulate solids in the leachate built up at the bottom of the SAMBR

eventually leading to the diffuser blocking. At MLTSS beyond 20 g/L, the TMP culminated at 850 mbar and the flux dropped to 0.5 LMH until the end of the experiment.

- The permeate COD cannot be lowered by decreasing the sparging rate. The main effect of low sparging rate (2LPM) was the enhanced rejection of the membrane and the significant flux drop while the permeate COD remained relatively constant. On the other hand, when the sparging rate was increased to 8-10 LPM, the permeate COD increased typically from 360 to 440 mg/L due to a better scouring of the membrane, which resulted in a thinner concentration polarization layer and a better diffusion through the biofilm. It is likely that the time required to observe the increase in permeate COD depends on the compaction of the cake layer. The increase in sparging rate allowed to recover the operation at 0 mbar TMP but not the initial flux.
- The permeate of the SAMBR was low in COD and relatively constant which promoted the growth of autotrophic bacteria in the subsequent AMBR so that 97.7% of the NH_4^+-N was removed at a maximum nitrogen loading rate of 0.18 kg NH_4^+-N/m^3.day. Complete nitrification was not observed in the AMBR as the nitrite concentration did not reach zero. This could be due to the low HRT (1 day) that could have overloaded the NOB such as *Nitrobacter*, or because of the lack of close distance between the nitrite producers and nitrite oxidisers which is required to obtain complete nitrification.
- Calcium in the leachate was found to precipitate in the AMBR because of the higher pH. A sample taken from the membrane consisted most likely of pure hydroxylapatite $Ca_5(PO_4)_3(OH)$ which had a needle shape, whereas the background of the precipitate consisted more of nodules of calcium carbonate with traces of manganese, iron, magnesium, aluminium, sulphur and sodium.
- The very high MW aromatic organics in the leachate fed to the SAMBRs were almost fully degraded in the bulk of the SAMBRs.

Moreover, their permeate was absolutely free of them which indicates a full rejection of these compounds by the membrane. The medium MW compounds in the range of 395 - 646 kDa were more likely to be rejected by the membrane while the MW ≤ 395 kDa were observed in the permeate. Regarding the evolution of the medium and low MW compounds over the 200 days, it can be stated that overall there was no build-up in the absorbance in any effluent, showing that these aromatics could be slowly degraded.

- Most of the recalcitrant compounds detected by GC-MS in the effluent of the HR were below 10 μg/L, while only two compounds (Bisphenol A and unidentified compounds with RI = 2096) were in the range 14 - 38 μg/L. The molecules o-hydroxybiphenyl and Bisphenol A were found to be non-biodegradable anaerobically because their areas increased in SAMBR1 and SAMBR2. The concentration of the former was in the range 11 – 22 μg/L, while the latter was in the range 21-44 μg/L. Moreover, as Bisphenol A was not found in the AMBR permeate, it can be stated that it was fully biodegraded aerobically. Some molecules were found to be non-biodegradable in an anaerobic environment, but could be slowly biodegraded in the AMBR such as 2,5-cyclohexadien - 1 - one,2,6 bis(1,1-dimethylethyl)- 4 -ethylidene- and Bis (2-ethylhexyl) phthalate. Nevertheless, the component N-phenethylbenzenesulfonamide were found to build-up in the AMBR and reached concentrations in the range 44-87 μg/L. Furthermore, new molecules appeared in the AMBR permeate such as Diisobutyl phthalate.

REFERENCES

Adani, F., F. Tambone, and A. Gotti. (2004). Biostabilization of municipal solid waste. *Waste Management* 24 (8):775-783.

Agdag, O. N., and D. T. Sponza. (2005). Anaerobic/aerobic treatment of municipal landfill leachate in sequential two-stage up-flow anaerobic sludge blanket reactor (UASB)/completely stirred tank reactor (CSTR) systems. *Process Biochemistry* 40 (2):895-902.

Agdag, O. N., and D. T. Sponza. (2007). Co-digestion of mixed industrial sludge with municipal solid wastes in anaerobic simulated landfilling bioreactors. *Journal of Hazardous Materials* 140 (1-2):75-85.

Ahn, Won-Young, Moon-Sun Kang, Seong-Keun Yim, and Kwang-Ho Choi. (2002). Advanced landfill leachate treatment using an integrated membrane process. *Desalination* 149 (1-3):109-114.

Akarsubasi, Alper T., Orhan Ince, Betul Kirdar, Nilgun A. Oz, Derin Orhon, Thomas P. Curtis, Ian M. Head, and Bahar K. Ince. (2005). Effect of wastewater composition on archaeal population diversity. *Water Research* 39 (8):1576-1584.

Akram, A. 2006. Stability and performance improvement of a Submerged Anaerobic Membrane Bioreactor (SAMBR) for wastewater treatment. PhD thesis, Department of Chemical Engineering and Chemical Technology, Imperial College London., London.

Anderson, G. K., and C. B. Saw. 1992. Leach-bed two-phase anaerobic digestion of municipal solid waste. Paper read at International Symposium on anaerobic digestion of solid waste, at Venice, Italy.

Anthonisen, A. C., R. C. Loehr, T. B. S. Prakasam, and E. G. Srinath. (1976). Inhibition of Nitrification by Ammonia and Nitrous-Acid. *Journal Water Pollution Control Federation* 48 (5):835-852.

APHA. (1999). *Standard Methods for the Examination of Water and Wastewater*. Edited by A. D. Eaton, L. S. Clesceri, A. E. Greenberg and M. A. H. Franson. Washington D.C: American Public Health Association.

Asakura, Hiroshi, Toshihiko Matsuto, and Nobutoshi Tanaka. (2004). Behavior of endocrine-disrupting chemicals in leachate from MSW landfill sites in Japan. *Waste Management* 24 (6):613-622.

Banks, C. J., and P. N. Humphreys. (1998). The anaerobic treatment of a ligno-cellulosic substrate offering little natural pH buffering capacity. *Water Science and Technology* 38 (4-5):29-35.

Barat, R., A. Bouzas, N. Marti, J. Ferrer, and A. Seco. (2008). Precipitation assessment in wastewater treatment plants operated for biological nutrient removal: A case study in Murcia, Spain. *Journal of Environmental Management* 90 (2):850-857.

Bensadoun, A., and D. Weinstein. (1976). Assay of Proteins in Presence of Interfering Materials. *Analytical Biochemistry* 70 (1):241-250.

Bilgili, M. S., A. Demir, and B. Ozkaya. (2007). Influence of leachate recirculation on aerobic and anaerobic decomposition of solid wastes. *Journal of Hazardous Materials* 143 (1-2):177-183.

Borzacconi, L., I. Lopez, M. Ohanian, and M. Vinas. (1999). Anaerobic-aerobic treatment of municipal solid waste leachate. *Environmental Technology* 20 (2):211-217.

Boyle, W. C., and R. K. Ham. (1974). Biological Treatability of Landfill Leachate. *Journal Water Pollution Control Federation* 46 (5):860-872.

Braber, K. (1995). Anaerobic digestion of municipal solid waste: A modern waste disposal option on the verge of breakthrough. *Biomass and Bioenergy* 9 (1-5):365-376.

Brockmann, M., and C. F. Seyfried. (1997). Sludge activity under the conditions of crossflow microfiltration. *Water Science and Technology* 35 (10):173-181.

Chang, J. E. (1989). Treatment of Landfill Leachate with an Upflow Anaerobic Reactor Combining a Sludge Bed and a Filter. *Water Science and Technology* 21 (4-5):133-143.

Chen, Ye, Jay J. Cheng, and Kurt S. Creamer. (2008). Inhibition of anaerobic digestion process: A review. *Bioresource Technology* 99 (10):4044-4064.

Cheng, S. S., and W. C. Chen. (1994). Organic-Carbon Supplement Influencing Performance of Biological Nitritification in a Fluidized-Bed Reactor. *Water Science and Technology* 30 (11):131-142.

Chesters, S. P. (2009). Innovations in the inhibition and cleaning of reverse osmosis membrane scaling and fouling. *Desalination* 238 (1-3):22-29.

Choi, J.-H., and H. Y. Ng. (2008). Effect of membrane type and material on performance of a submerged membrane bioreactor. *Chemosphere* 71 (5):853-859.

Choo, K. H., and C. H. Lee. (1996). Effect of anaerobic digestion broth composition on membrane permeability. *Water Science and Technology* 34 (9):173-179.

Cook, E. N., and E. G. Foree. (1974). Aerobic Biostabilization of Sanitary Landfill Leachate. *Journal Water Pollution Control Federation* 46 (2):380-392.

DEFRA. 2005. Municipal Waste Management Survey 2003/2004, edited by Crown: Department for Environment, Food and Rural Affairs.

DEFRA. 2008. Municipal Waste Management Statistics. In *Provisional Quarter 1 2008/09*: Department for Environment, Food and Rural Affairs.

Dierick, B. 2006. Marketing Engineer OWS Dranco. Personnal communication.

Doyle, J. D., and S. A. Parsons. (2002). Struvite formation, control and recovery. *Water Research* 36 (16):3925-3940.

Dubois, M., K. A. Gilles, J. K. Hamilton, P. A. Rebers, and F. Smith. (1956). Colorimetric Method for Determination of Sugars and Related Substances. *Analytical Chemistry* 28:350-356.

Elmaleh, S., and L. Abdelmoumni. (1997). Cross-flow filtration of an anaerobic methanogenic suspension. *Journal of Membrane Science* 131 (1-2):261-274.

Feijoo, G., M. Soto, R. Mendez, and J. M. Lema. (1995). Sodium Inhibition in the Anaerobic-Digestion Process - Antagonism and Adaptation Phenomena. *Enzyme and Microbial Technology* 17 (2):180-188.

Fernandez, N., E. E. Diaz, R. Amils, and J. L. Sanz. (2008). Analysis of microbial community during biofilm development in an anaerobic wastewater treatment reactor. *Microbial Ecology* 56 (1):121-132.

Fricke, K., H. Santen, R. Wallmann, A. Huttner, and N. Dichtl. (2007). Operating problems in anaerobic digestion plants resulting from nitrogen in MSW. *Waste Management* 27 (1):30-43.

Fromme, Hermann, Thomas Kuchler, Thomas Otto, Konstanze Pilz, Josef Muller, and Andrea Wenzel. (2002). Occurrence of phthalates and bisphenol A and F in the environment. *Water Research* 36 (6):1429-1438.

Gallert, C., A. Henning, and J. Winter. (2003). Scale-up of anaerobic digestion of the biowaste fraction from domestic wastes. *Water Research* 37 (6):1433-1441.

Gray, N. D., I. P. Miskin, O. Kornilova, T. P. Curtis, and I. M. Head. (2002). Occurrence and activity of Archaea in aerated activated sludge wastewater treatment plants. *Environmental Microbiology* 4 (3):158-168.

Hanaki, K., C. Wantawin, and S. Ohgaki. (1990). Effects of the Activity of Heterotrophs on Nitrification in a Suspended-Growth Reactor. *Water Research* 24 (3):289-296.

Hao, Y.-J., W.-X. Wu, S.-W. Wu, H. Sun, and Y.-X. Chen. (2008). Municipal solid waste decomposition under oversaturated condition in comparison with leachate recirculation. *Process Biochemistry* 43 (1):108-112.

Harada, H., K. Momonoi, S. Yamazaki, and S. Takizawa. (1995). Application of anaerobic-UF membrane reactor for treatment of a wastewater containing high strength particulate organics. *Water Science and Technology* 30 (12):307-319.

Hartmann, H., and B. K. Ahring. (2005). Anaerobic digestion of the organic fraction of municipal solid waste: Influence of co-digestion with manure. *Water Research* 39 (8):1543-1552.

He, P. J., F. Lu, L. M. Shao, X. J. Pan, and D. J. Lee. (2006). Effect of alkali metal cation on the anaerobic hydrolysis and acidogenesis of vegetable waste. *Environmental Technology* 27 (3):317-327.

He, R., D. S. Shen, J. Q. Wang, Y. H. He, and Y. M. Zhu. (2005). Biological degradation of MSW in a methanogenic reactor using treated leachate recirculation. *Process Biochemistry* 40 (12):3660-3666.

Held, C., M. Wellacher, K.-H. Robra, and G. M. Gubitz. (2002). Two-stage anaerobic fermentation of organic waste in CSTR and UFAF-reactors. *Bioresource Technology* 81 (1):19-24.

Hendriks, A. T. W. M., and G. Zeeman. (2009). Pretreatments to enhance the digestibility of lignocellulosic biomass. *Bioresource Technology* 100 (1):10-18.

Henry, J. G., D. Prasad, and H. Young. (1987). Removal of organics from leachates by anaerobic filter. *Water Research* 21 (11):1395-1399.

Hoilijoki, T. H., R. H. Kettunen, and J. A. Rintala. (2000). Nitrification of anaerobically pretreated municipal landfill leachate at low temperature. *Water Research* 34 (5):1435-1446.

Hong, S. P., T. H. Bae, T. M. Tak, S. Hong, and A. Randall. (2002). Fouling control in activated sludge submerged hollow fiber membrane bioreactors. *Desalination* 143 (3):219-228.

Horn, Owen, Sandro Nalli, David Cooper, and Jim Nicell. (2004). Plasticizer metabolites in the environment. *Water Research* 38 (17):3693-3698.

Hu, Y.-C. 2004. Submerged Anaerobic Membrane Bioreactor (SAMBR) for Wastewater treatment. PhD thesis, Department of Chemical Engineering, Imperial College London., London.

Hulshoff Pol, L. W., W. J. de Zeeuw, C. T. M. Velzeboer, and G. Lettinga. (1983). Granulation in UASB reactors. *Water Science and Technology* 15:291-304.

Im, Jeong-hoon, Hae-jin Woo, Myung-won Choi, Ki-back Han, and Chang-won Kim. (2001). Simultaneous organic and nitrogen removal from municipal landfill leachate using an anaerobic-aerobic system. *Water Research* 35 (10):2403-2410.

Ito, A., M. Maekawa, T. Tsutsumi, F. Ikazaki, and T. Tateishi. (1997). Solubility product of OH-carbonated hydroxyapatite. *Journal of Biomedical Materials Research* 36 (4):522-528.

Jackson-Moss, C. A., J. R. Duncan, and D. R. Cooper. (1989). The effect of calcium on anaerobic digestion. *Biotechnol. Lett.* 11 (3):219-224.

Jokela, J. P. Y., R. H. Kettunen, K. M. Sormunen, and J. A. Rintala. (2002). Biological nitrogen removal from municipal landfill leachate: low-cost nitrification in biofilters and laboratory scale \textit{in-situ} denitrification. *Water Research* 36 (16):4079-4087.

Kayhanian, M., and D. Rich. (1996). Pilot-scale high solids thermophilic anaerobic digetion of municipal solid waste with an emphasis on nutrient requirements. *Fuel and Energy Abstracts* 37 (3):237-237.

Kennedy, K. J., M. F. Hamoda, and S. G. Guiot. (1988). Anaerobic Treatment of Leachate Using Fixed Film and Sludge Bed Systems. *Journal Water Pollution Control Federation* 60 (9):1675-1683.

Kim, M., Y. H. Ahn, and R. E. Speece. (2002). Comparative process stability and efficiency of anaerobic digestion; mesophilic vs. thermophilic. *Water Research* 36 (17):4369-4385.

Kowalchuk, G. A., P. L. E. Bodelier, G. H. J. Heilig, J. R. Stephen, and H. J. Laanbroek. (1998). Community analysis of ammonia-oxidising bacteria, in relation to oxygen availability in soils and root-oxygenated sediments, using PCR, DGGE and oligonucleotide probe hybridisation. *Fems Microbiology Ecology* 27 (4):339-350.

Kugelman, I. J., and P. L. McCarty. (1964). Cation toxicity and stimulation in anaerobic waste treatment. *J. Water Pollut. Control Fed.* 37:97-116.

Laitinen, N., A. Luonsi, and J. Vilen. (2006). Landfill leachate treatment with sequencing batch reactor and membrane bioreactor. *Desalination International Congress on Membranes and Membrane Processes* 191 (1-3):86-91.

Lay, J. J., T. Noike, G. Endo, and S. Ishimoto. (1997). Analysis of environmental factors affecting methane production from high-solids organic waste. *Water science and technology* 36 (6-7):493-500.

Le Corre, K. S., E. Valsami-Jones, P. Hobbs, and S. A. Parsons. (2005). Impact of calcium on struvite crystal size, shape and purity. *Journal of Crystal Growth* 283 (3-4):514-522.

Lee, D. H., S. K. Behera, J. W. Kim, and H. S. Park. (2009). Methane production potential of leachate generated from Korean food waste recycling facilities: A lab-scale study. *Waste Management* 29 (2):876-882.

Lee, J. P., S. C. Park, and S. W. Rhee. (1997). The effect of ammonia and sodium chloride on the anaerobic digestion of food waste. *Journal of Korean Society of Environmental Engineers* 19 (9):1185-1192.

Li, H., A. G. Fane, H. G. L. Coster, and S. Vigneswaran. (2003). Observation of deposition and removal behaviour of submicron bacteria on the membrane surface during crossflow microfiltration. *Journal of Membrane Science* 217 (1-2):29-41.

Liao, Wei, Yan Liu, Chuanbin Liu, and Shulin Chen. (2004). Optimizing dilute acid hydrolysis of hemicellulose in a nitrogen-rich cellulosic material-dairy manure. *Bioresource Technology* 94 (1):33-41.

Lin, Y. P., and P. C. Singer. (2005). Inhibition of calcite crystal growth by polyphosphates. *Water Research* 39 (19):4835-4843.

Liu, S. T., F. L. Yang, Z. Gong, F. G. Meng, H. H. Chen, Y. Xue, and K. J. Furukawa. (2008). Application of anaerobic ammonium-oxidizing consortium to achieve completely autotrophic ammonium and sulfate removal. *Bioresource Technology* 99 (15):6817-6825.

Liu, Yitai, and David R. Boone. (1991). Effects of salinity on methanogenic decomposition. *Bioresource Technology* 35 (3):271-273.

Lopes, W. S., V. D. Leite, and S. Prasad. (2004). Influence of inoculum on performance of anaerobic reactors for treating municipal solid waste. *Bioresource Technology* 94 (3):261-266.

MacRae, Jean D., and Kenneth J. Hall. (1998). Biodegradation of polycyclic aromatic hydrocarbons (PAH) in marine sediment under denitrifying conditions. *Water Science and Technology* 38 (11):177-185.

Mahmoud, N. 2002. Anaerobic pre-treatment of sewage under low temperature (15\degre C) conditions in an integrated UASB-Digester system. PhD Thesis, Department of Environmental Technology, Wageningen University, The Netherlands., Wageningen.

Mamais, D., P. A. Pitt, Y. W. Cheng, J. Loiacono, and D. Jenkins. (1994). Determination of Ferric-Chloride Dose to Control Struvite Precipitation in Anaerobic Sludge Digesters. *Water Environment Research* 66 (7):912-918.

Mansouri, L., L. Bousselmi, and A. Ghrabi. (2007). Degradation of recalcitrant organic contaminants by solar photocatalysis. *Water Science and Technology* 55 (12):119-125.

Maris, P. J., D. W. Harrington, and G. L. Chismon. (1984). Leachate Treatment with Particular Reference to Aerated Lagoons. *Water Pollution Control* 83 (4):521-538.

Mathiesen. 1989. Ca and/or Mg soap solution in biogas production. WO8900548 (Patent).

McCarty, P. L. (1964). Anaerobic waste treatment fundamentals. *Public works* 95 (9):107-112.

McCarty, P. L., and R. McKinney. (1961). Salt toxicity in anaerobic digestion. *J. Water Pollut. Control Fed.* 33:399-415.

Monson, K. D., S. R. Esteves, A. J. Guwy, and R. M. Dinsdale. (2007). *Anaerobic digestion of biodegradable municipal solid waste: A review*: University of Glamorgan.

Montastruc, L., C. Azzaro-Pantel, B. Biscans, M. Cabassud, and S. Domenech. (2003). A thermochemical approach for calcium phosphate precipitation modeling in a pellet reactor. *Chemical Engineering Journal* 94 (1):41-50.

Nalli, S., D. G. Cooper, and J. A. Nicell. (2003). Biodegradation of plasticizers by \textit{Rhodococcus rhodochrous}. *Biodegradation* 13 (5):343-352.

Nopharatana, A., P. C. Pullammanappallil, and W. P. Clarke. (2006). Kinetics and dynamic modelling of batch anaerobic digestion of municipal solid waste in a stirred reactor. *Waste Management* 27 (5):595-603.

Ortega-Clemente, Alfredo, S. Caffarel-Mendez, M. T. Ponce-Noyola, J. Barrera-Cortes, and Hector M. Poggi-Varaldo. (2008). Fungal post-treatment of pulp mill effluents for the removal of recalcitrant pollutants. *Bioresource Technology* 100 (6):1885-1894.

Owen, W. F., D. C. Stuckey, Jr. Healy, J. B., L. Y. Young, and P. L. McCarty. (1979). Bioassay for monitoring biochemical methane potential and anaerobic toxicity. *Water Research* 13 (6):485-492.

Perrin, D. D. (1982). *Ionisation Constants of inorganic acids and bases in aqueous solutions*. Edited by I. U. o. P. a. A. c. (IUPAC). 2nd edition ed. UK: Pergamon Press.

Phillips, Paul S., Richard Barnes, Margaret P. Bates, and Thomas Coskeran. (2005). A critical appraisal of a UK county waste minimisation programme: The requirement for regional facilitated development of industrial symbiosis/ecology. *Resources, Conservation and Recycling* 46 (3):242-264.

Plant, L. J., and W. A. House. (2002). Precipitation of calcite in the presence of inorganic phosphate. *Colloids and Surfaces A: Physicochemical and Engineering Aspects* 203 (1-3):143-153.

Prosser, J. I. (2007). The ecology of Nitrifying bacteria. In *Biology of the Nitrogen Cycle*, edited by H. Bothe, S. J. Ferguson and W. E. Newton. Amsterdam, The Netherlands: Elsevier.

Ramphao, M., G. A. Ekama, M. T. Lakay, H. Mafungwa, and M. C. Wentzel. 2004. The performance and kinetics of biological nitrogen and phosphorus removal with ultra-filtration membranes for solid-liquid separation. UCT MSc Thesis, Department of Civil Engineering, University of Cape Town, Cape Town.

Ritchie, G. A. F., and D. J. Nicholas. (1972). Identification of Sources of Nitrous-Oxide Produced by Oxidative and Reductive Processes in *Nitrosomonas europaea*. *Biochemical Journal* 126 (5):1181-1191.

Rittman, B. E., and P. L. McCarty. (2001). *Environmental biotechnology: Principles and applications*. London: McGraw-Hill Int. Editions.

Robinson, A. H. 2005. Landfill leachate treatment. Paper read at 5th International Conference on Membrane Bioreactors, at Cranfield University, UK.

Robinson, H. D., and P. J. Maris. (1985). The treatment of leachates from domestic waste in landfill sites. *Journal of the Water Pollution Control Association* 57 (1):30-38.

Ross, W. R., J. P. Barnard, J. Leroux, and H. A. Devilliers. (1990). Application of Ultrafiltration Membranes for Solids - Liquid Separation in Anaerobic-Digestion Systems - the ADUF Process. *Water S.A.* 16 (2):85-91.

Sanders, W. T. M. 2001. Anaerobic hydrolysis during digestion of complex substrates. PhD Thesis, Department of Environmental Technology, Wageningen University, The Netherlands., Wageningen.

Scaglia, B., and F. Adani. (2008). An index for quantifying the aerobic reactivity of municipal solid wastes and derived waste products. *Science of The Total Environment* 394 (1):183-191.

Schmidt, J. E., and B. K. Ahring. (1993). Effects of magnesium on thermophilic acetate-degrading granules in upflow anaerobic sludge blanket (UASB) reactors. *Enzyme and Microbial Technology* 15 (4):304-310.

Schwarzbauer, Jan, Sabine Heim, Sabine Brinker, and Ralf Littke. (2002). Occurrence and alteration of organic contaminants in seepage and leakage water from a waste deposit landfill. *Water Research* 36 (9):2275-2287.

Seely, O. 2008. Selected Solubility Products and Formation Constants at 25°C. California State University, Dominguez Hills: http://www.csudh.edu/oliver/chemdata/data-ksp.htm.

Shin, H. S., M. J. Moon, Y. C. Song, and B. U. Bae. (1993). A study on anaerobic treatability of food waste using biodegradability test. *Journal of Korean Society of Waste Management* 10 (1):35-42.

Shin, H. S., Y. C. Song, and B. C. Paik. (1995). Inhibtory effects of sodium ions on the anaerobic degradation of food waste. *Journal of Koren Solid Wastes Engineering Society* 12 (1):50-57.

Shin, J.-H., S.-M. Lee, J.-Y. Jung, Y.-C. Chung, and S.-H. Noh. (2005). Enhanced COD and nitrogen removals for the treatment of swine wastewater by combining submerged membrane bioreactor (MBR) and anaerobic upflow bed filter (AUBF) reactor. *Process Biochemistry* 40 (12):3769-3776.

Shpiner, R. 2007. Treatment of Produced Water by Waste Stabilisation Ponds. PhD thesis, Chemical Engineering and Chemical Technology, Imperial College London., London.

Sierra-Alvarez, Reyes, Jolan Harbrecht, Sjon Kortekaas, and Gatze Lettinga. (1990). The continuous anaerobic treatment of pulping wastewaters. *Journal of Fermentation and Bioengineering* 70 (2):119-127.

Soto, M., R. Mendez, and J. M. Lema. (1993). Sodium Inhibition and Sulfate Reduction in the Anaerobic Treatment of Mussel Processing Wastewaters. *Journal of Chemical Technology and Biotechnology* 58 (1):1-7.

Speece, R. E. (1996). *Anaerobic Biotechnology for Industrial Wastewaters*. Nashville, Tennessee, USA: Archae Press.

Spivack, J., T. K. Leib, and J. H. Lobos. (1994). Novel Pathway for Bacterial Metabolism of Bisphenol-a - Rearrangements and Stilbene Cleavage in Bisphenol-a Metabolism. *Journal of Biological Chemistry* 269 (10):7323-7329.

Stephen, J. R., G. A. Kowalchuk, M. A. V. Bruns, A. E. McCaig, C. J. Phillips, T. M. Embley, and J. I. Prosser. (1998). Analysis of beta-subgroup proteobacterial ammonia oxidizer populations in soil by denaturing gradient gel electrophoresis analysis and hierarchical phylogenetic probing. *Applied and Environmental Microbiology* 64 (8):2958-2965.

Trosch, W, and V. Niemann. 1999. Biological waste treatment using the thermophilic Schwarting-Uhde process. Paper read at II Int. Symp. on Anaerobic Digestion of Solid Waste, 15-17 June, at Barcelona.

Trzcinski, A, P., and D. C. Stuckey. (2009a). Reproduced from "Anaerobic digestion of the organic fraction of municipal solid waste in a two-stage membrane process." *Water Science & Technology* 60 (8):1965-1978, with permission from the copyright holders, IWA Publishing.

Trzcinski, A. P., and D. C. Stuckey. (2009b). Reprinted from "Continuous treatment of the Organic Fraction of Municipal Solid Waste in an anaerobic two-stage membrane process with liquid recycle." *Water Research* 43 (9):2449-2462, Copyright (2009), with permission from Elsevier.

Trzcinski, Antoine P., and David C. Stuckey. (2016). Reprinted from "Inorganic fouling of an anaerobic membrane bioreactor treating leachate from the organic fraction of municipal solid waste (OFMSW) and a polishing aerobic membrane bioreactor." *Bioresource Technology* 204:17-25, Copyright (2016), with permission from Elsevier.

Uloth, V. C., and D. S. Mavinic. (1977). Aerobic Bio-Treatment of a High-Strength Leachate. *Journal of the Environmental Engineering Division-Asce* 103 (4):647-661.

Van Den Doole, H., and P.D. Kratz. (1963). A generalization of the retention index system including linear temperature programmed gas-liquid partition chromatography *J. Chromatography* 11:463-471.

Vandevivere, P., L. De Baere, and W. Verstraete. 2003. Types of anaerobic digesters for solid wastes. Paper read at biomethanization of the Organic Fraction of Municipal Solid Wastes, at London.

Vieitez, E. R., J. Mosquera, and S. Ghosh. (2000). Kinetics of accelerated solid-state fermentation of organic-rich municipal solid waste. *Water Science and Technology* 41 (3):231-238.

Vyrides, I., and D. C. Stuckey. (2009). Saline sewage treatment using a submerged anaerobic membrane reactor (SAMBR): Effects of activated carbon addition and biogas-sparging time. *Water Research* 43 (4):933-942.

Wang, Feng, Si-qing Xia, Yi Liu, Xue-song Chen, and Jun Zhang. (2007). Community analysis of ammonia and nitrite oxidizers in start-up of aerobic granular sludge reactor. *Journal of Environmental Sciences* 19 (8):996-1002.

Weast, R. C. (1963). *Handbook of chemistry and physics: a ready-reference book of chemical and physical data*. 44th ed. Cleveland, Ohio: CRC Press.

Wellinger, A., C. Widmer, and P. Schalk. 1999. Percolation - a new process to treat MSW. Paper read at Second International Symposium on Anaerobic Digestion of Solid Wastes, at Barcelona.

White, T. J., T. Bruns, S. Lee, and J. Taylor. (1990). Amplification and direct sequencing of fungal ribosomal RNA genes for phylogenetics. In *PCR Protocols: A Guide to Methods and Applications* edited by M. A. Innis, D. H. Gelfand, J. J. Sninsky and T. J. White. New York: Academic Press.

Wintgens, T., M. Gallenkernper, and T. Melin. (2004). Removal of endocrine disrupting compounds with membrane processes in wastewater treatment and reuse. *Water Science and Technology* 50 (5):1-8.

Wittebolle, L., H. Vervaeren, W. Verstraete, and N. Boon. (2008). Quantifying community dynamics of nitrifiers in functionally stable reactors. *Applied and Environmental Microbiology* 74 (1):286-293.

Xu, Yiping, Yiqi Zhou, Donghong Wang, Shaohua Chen, Junxin Liu, and Zijian Wang. (2008). Occurrence and removal of organic micropollutants in the treatment of landfill leachate by combined anaerobic-membrane bioreactor technology. *Journal of Environmental Sciences* 20 (11):1281-1287.

Yamamoto, K., M. Hiasa, T. Mahmood, and T. Matsuo. (1989). Direct Solid-Liquid Separation Using Hollow Fiber Membrane in an Activated-Sludge Aeration Tank. *Water Science and Technology* 21 (4-5):43-54.

Yamamoto, Takashi, Akio Yasuhara, Hiroaki Shiraishi, and Osami Nakasugi. (2001). Bisphenol A in hazardous waste landfill leachates. *Chemosphere* 42 (4):415-418.

Yasuhara, Akio, H. Shiraishi, M. Nishikawa, T. Yamamoto, and O. Nakasugi. (1999). Organic components in leachates from hazardous waste disposal sites. *Waste management and research* 17 (3):186.

Zhang, R., H. M. El-Mashad, K. Hartman, F. Wang, G. Liu, C. Choate, and P. Gamble. (2007). Characterization of food waste as feedstock for anaerobic digestion. *Bioresource Technology* 98:929-935.

In: Leachates
Editor: Maria R. Mattox

ISBN: 978-1-53612-508-5

Chapter 2

CERAMIC RASCHIG RINGS–IMPROVING REMOVAL OF ORGANIC COMPOUNDS FROM LANDFILL LEACHATE BY PEOZONE (O_3/H_2O_2)

***Huu Tap Van*[1,*] *and Van Tuyen Trinh*[2]**
[1]Faculty of Environment and Earth Science,
Thai Nguyen University of Sciences,
Thai Nguyen city, Vietnam
[2]Institute of Environmental Technology,
Vietnam Academy of Science and Technology,
Ha Noi, Vietnam

ABSTRACT

For improving the treatment of landfill leachate by combination O_3/H_2O_2 and Ceramic Raschig Rings (CRR) with different surface areas were added to enhance the mass transfer of ozone into liquid phase. To determine optimal conditions of reaction, pH and reaction time were also studied. The optimal pH ranges of 8-9, optimal reaction time of 80 min were identified in this research. CRR contribute to the significant

[*]Corresponding Author Email : vanhuutap@gmail.com.

improvement of the efficiency of landfill leachate treatment by O_3/H_2O_2 systems. With added CRR of 728m^2/m^3 surface area, color, COD and TOC removal were increased in comparison with experiments without CRR was 8%, 14% and 9%, respectively. In this condition, the ozone utilization efficiency was also higher than that in experiments without CRR. The content of O_3 was also identified using 3.441 kg O_3/kg COD.

Keywords: Ozone, Ceramic Raschig Ring, Color, Landfill leachate, O_3/H_2O_2

INTRODUCTION

Landfilling is still a popular method to eliminate solid waste around the world, as it is the cheapest method in comparison with incineration and composting (Sumanaweera 2004). Generation of leachate is a significant problem of landfill; it is commonly generated as a result of water from precipitation, surface run-off, and infiltration or intrusion of groundwater percolating through the landfill (Wu et al. 2003). Leachate produced from landfills is highly polluted wastewater which might contain the large amount of organic and inorganic compounds, such as humic matter, ammonia nitrogen, heavy metals, chlorinated organics and inorganic salts (Zhou et al. 2009). Untreated leachate can pollute ground water, surface water and contribute to contaminate soil.

Leachate channeling (Combined treatment with domestic sewage, Recycling and lagooning), Biological Processes (Aerobic and Anaerobic) and Chemical/Physical Treatment (Chemical precipitation, Chemical oxidation, Adsorption, membrane technology and stripping of NH_3) are some of the leachate treatment methods (Sumanaweera 2004). Many researchers recognized that simple biological or chemical treatment methods were not sufficient for treatment of landfill leachate (Ratanatamskul et al. 2009). The application of Advanced Oxidation Processes (AOPs) have been considered to enhance significantly landfill leachate treatment by improving BOD/COD ratio with the use of O_3, O_3/UV, O_3/H_2O_2, O_3/H_2O_2/UV and Fenton processes (Haapea et al. 2002,

Ratanatamskul et al. 2009, Oliveira et al. 2010, Zhou et al. 2009, Schulte et al. 1995, Wang et al. 2004,). Combined action of activated carbon and ozone can improve the treatment of municipal landfill leachate by reducing COD, color and the transition metals (Cataldo et al. 2013).

Although ozone is a powerful oxidant, dissolving ability from gas phase into liquid phase is limited. When ozone is added to water, it needs to be transferred from the gas phase mixture into the liquid phase. The necessary condition for that is the surface between the two phases enables phase transfer (Alwasit et al. 2009).

In Viet Nam, most solid waste from urban and industrial areas were buried in landfill sites but generated leachate was not treated thoroughly. Landfill sites generate the large amounts of leachate containing highly toxic and non-biodegradable organic compounds. Landfill leachate treatment has recently been of concern in. Many systems were set up to treat landfill leachate due to local pressures (Khanh 2009). Researches of landfill leachate are limited in Viet Nam, the biological method of landfill leachate treatment is the most popular. This method is insufficient to treat non-biological organic compounds from landfill leachate.

The goal of this research is addition of Ceramic Raschig Ring as a new agent to enhance mass transfer of ozone from gas phase into liquid phase (landfill leachate) for treatment of organic compounds. As a single process, oxidation with ozone is not considered as very effective due to the complexity of leachate composition. H_2O_2 can enhance ozone decomposition to yeild $^{\circ}OH$ (Li et al. 2010). Thus, the combination of ozone, hydrogen peroxide and ceramic raschig ring (O_3/H_2O_2/CRR) is applied to treat organic compounds from landfill leachate in this research. Ozone concentration in reactor with ceramic raschig ring can be higher than in traditional reactor. $^{\circ}OH$ concentration can be also enhanced by the presence of H_2O_2. Therefore, ozone alone is not considered in this study.

Ceramic Rasching rings have excellent acid and heat resistance. It can resist corrosion of various inorganic and organic acids and organic solvents except hyhro fluoric acid, and can be used in high or low temperature conditions. Therefore, the application range is very wide. It can be used in chemical industry, metallurgy industry, coal gas industry and oxygen producing industry, etc. They are also used for devices where gas and liquid are placed in contact for purposes of gas absorption, stripping or chemical reaction, and as a support for biofilms in biological reactors. In this study, they are used to improve absorption of O_3 into landfill leachate from gas phase.

Ceramic raschig rings (CRR) have mechanical strength and specific surface that are many times higher than that of other conventional materials. With main chemical composition of SiO_2 and Al_2O_3, CRR have abilities to withstand high temperatures up to 800°C and are sustainable in high acidic or base medium. Chemical composition: >60% SiO_2, 17-23% Al_2O_3, <1% Fe_2O_3, 1-2% CaO_2, 2-4% K_2O + Na_2O. It is shown in table 1.

Materials and Methods

Landfill Leachate

Nam Son landfill site has been in operation since 1999 and located in Soc Son, Ha Noi city, Viet Nam. The total area of landfill is 54 hectares, divided into 9 buried boxes and about 2,000 tons of municipal solid waste is disposed daily. Landfill leachate is generated about 350-400 m^3/day. Characteristics of landfill leachate are shown in Table 2. It was a dark black alkaline mixture. The average BOD_5/COD ratio was below 0.40.

Table1. Features of Ceramic Raschig Ring (CRR)

Internal diameter (mm)	DxHxδ (mm)	Specific surface (m^2/m^3)	Porosity (%)	Quantity (rings/m^3)	Bulk density (kg/m^3)
2	2 x 10 x 3	728	44	2,880,000	1,215
4	4 x 12 x 2.5	539	54	910,000	1,156
6	6 x 10 x 3	356	59	1,042	931

Note: D: Internal diameter (mm), H: Hight (mm), δ: Thickness of wall of ring (mm).

Table 2. Characteristics of landfill leachate

Parameters	Unit	Value	Average
pH	-	7.87 - 8.52	8.20
Color	Pt-Co	2,025 – 3,050	2,538
COD	mg/l	2,178 – 3,717	2,948
TOC	mg/l	955 – 1,306	1.131
Total N	mg/l	724 – 1,806	1,265
NH_4^+	mg/l	670 – 1,668	1,169
Total P	mg/l	4.28 – 6.30	5.29
Cl^-	mg/l	1724 – 2,469	2,097
CO_3^{2-}	mg/l	198 - 234	216
HCO_3^-	mg/l	9,842 – 15,625	12,734
SO_4^{2-}	mg/l	99 - 282	191

Ammonia, color and chloride in the leachate are high, thus, it is difficult for treating.

Analytical Methods

Leachate was characterized before and after ozonation, through analyses of COD, TOC, Color and O_3 concentration were determined by standard method (APHA 2012).

Experimental Apparatus

Figure 1 shows our experiment setup. O_3/H_2O_2/CRR experiments consist of two reactors: one without CRR and one containing CRR (Figure 1a and 1b, respectively). The reactor was composed of buddle column to contact ozone gas with landfill leachate. It was acylindrical heat-resistant mica, the internal diameter of column was 54mm, height of 1,000mm and thickness of 3mm.

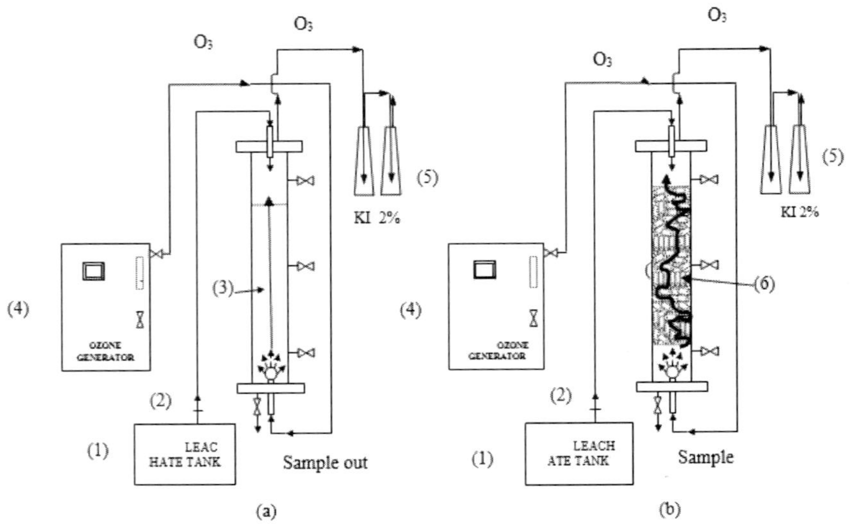

Figure 1. Experiment systems of landfill leachate (a) Without CRR, (b) With CRR. (1) Tank containing landfill leachate, (2) Landfill leachate pump, (3) Reactor, (4) Ozone generator, (5) Ozone absorption device, (6) Ceramic raschig ring.

Ozone was generated by Ozone generator (Lino Lin 4.10L), capable of producing up to 10 gO_3/h. The ozone was fed into the reactor through the bottom.

***.O_3/H_2O_2 Experiment**

With the presence of O_3 and H_2O_2, organic substances will be oxidazed in two methods: the first way is the direct ozonation of organic substances and the second way is oxidation of organic substances through hydroxyl radical (^oOH) (Gottschalk et al. 2010). ^oOH is generated by reaction:

$$H_2O_2 + 2O_3 \rightarrow 2OH^{\circ} + 3O_2 \quad (1)$$

The effect of initial pH was studied by adjusting the pH of the leachate with NaOH. In O_3/H_2O_2 experiments, the required amount of H_2O_2 was injected in a single-step to the reactor at time zero. Landfill leachate is pumped into the reactor with a volume of 1000ml/batch. Then an aeration containing ozone is introducted in a specific period of time, at an air flow of 7

L/min equal to O_3 capacity of 2.882 g/h, H_2O_2 solution 30% is used for this study. According to reaction equation (1), 2 mole O_3 react with 1 mole H_2O_2, thus, a mole ratio of O_3/H_2O_2: 0.8 is used for this research, the ratio is applied for the inlet mole of O_3 from gas phase. Residual ozone led into the device is absorbed by KI 2% solution.

***. O_3/H_2O_2/CRR Experiments**

The experiments in combining O_3/H_2O_2/CRR were made after determining optimal conditions of pH and reaction time in experiment systems with reactor without CRR. The aim of experiments is to assess the influence of the specific surface of CRR and O_3/H_2O_2 to the treatment capacity of landfill leachate.

The experiment is performed with specific surface values of CRR of 0; 356; 539 and 728m^2/m^3, respectively. The volume of landfill leachate and airflow of ozone generator are unchanged when comparing with the experiment without CRR.

Landfill leachate is pre-treated by flocculation-coagulation with PAC of 1,500 mg/l to remove suspended solids (SS) and the small amount of organic substances. All experiments were carried out at room temperature (23 ± 5°C).

RESULTS AND DISCUSSION

Effect of Initial pH

In O_3/H_2O_2 experiments without CRR, fixed processing time as 60 min. Each mode of experiments was performed 3 times, labeled Experiment 1, Experiment 2 and Experiment 3. After coagulation, concentrations of color, COD and TOC are 1,995 – 2,100 Pt-Co, 1,672 – 2,712 mg/l, 853 – 1,066 mg/l, respectively.

The effect of pH was investigated at 5; 6; 7; 8; 9 and 10. The effect of initial pH on degradation of the organic compounds from landfill leachate by O_3/H_2O_2 is given in figure 2.

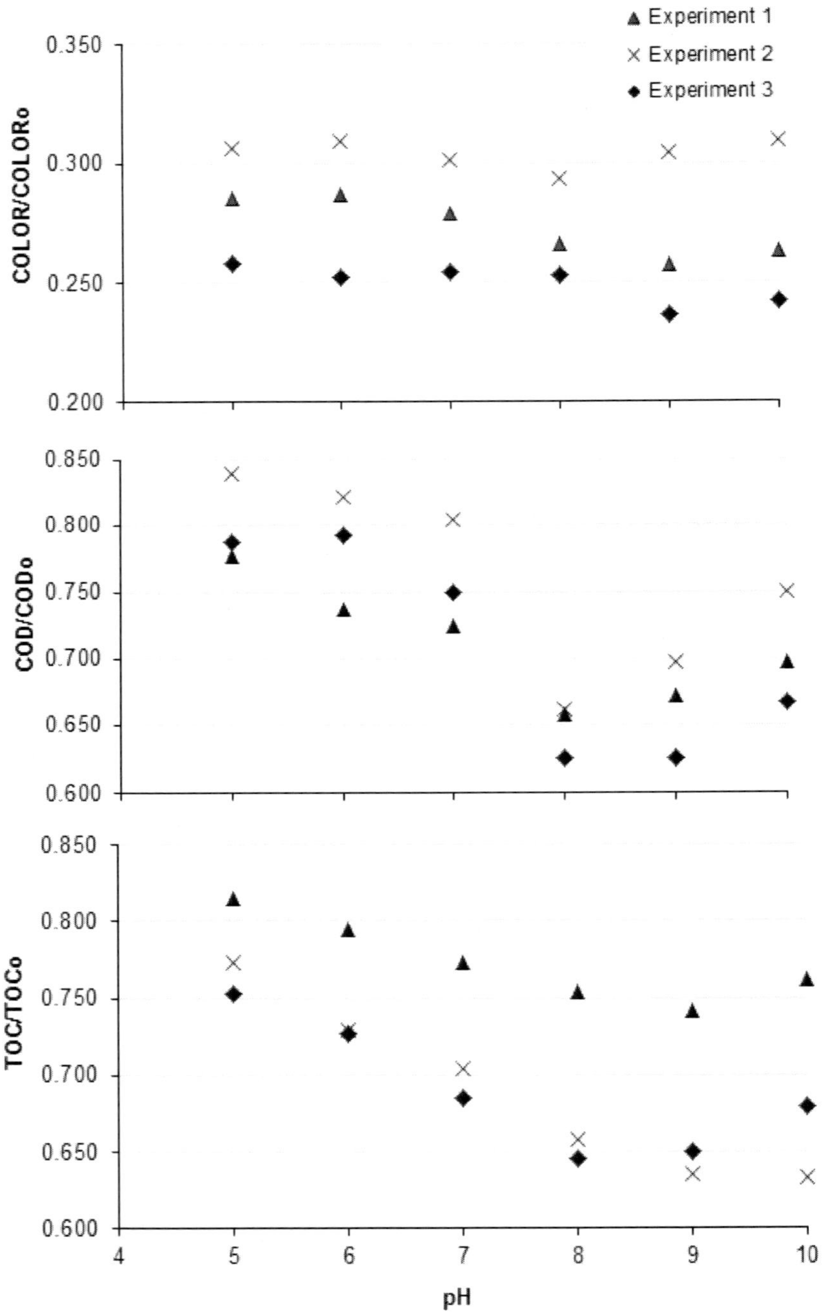

Figure 2. The effect of initial pH on Color, COD and TOC of landfill leachate treatment by O_3/H_2O_2 system without CRR, reaction time = 60 min.

Color, COD and TOC removal efficiencies increased as the pH range of 7-9 raised because under basic pH higher, the number of ozone molecules were decomposed to generate OH^- and ^oOH radicals and more powerful hydroxyl oxidation dominates. They decreased at pH 10. In pH range 5-7, color and COD removal increased slowly. With 60 min of reaction, the treatment efficiencies were not high, after flocculation, the treating performance of Color, COD and TOC achieved 68 - 76%, 16 - 39% and 15 - 42%, respectively. The highest efficiencies of COD and TOC were observed at initial pH of 8 and 9, was about 30-39% (reduced to 1,100 – 1,695 mg/l) and 20-42% (reduced to 613 – 693 mg/l), respectively.

At low pH, organic compounds were mainly oxidized by O_3 ($M + O_3$) molecules with slow reaction rate, typically being in the range of $k_D = 1.0 - 10^3\ M^{-1}s^{-1}$ (Gottschalk et al. 2010), the efficiency was usually lower. When increasing pH, the decay of ozone to form secondary oxidants such as hydroxyl radical (^oOH), the oxidation reactions mainly were occurred by hydroxyl radical.

$$O_3 + H_2O_2 \rightarrow 2^\circ HO + O_2 \qquad (2)$$

Therefore, at high pH, there was the significant effect of ozone oxidative capacity via hydroxyl radical (^oOH). However, at pH 10, the concentrations of color, COD and TOC tended to increase gradually after ozonation with combining O_3/H_2O_2. Oxidation through the formation of ^oOH radical was limited by the presence of ozone-resistant compounds or ^oOH radical scavengers (Kurniawan et al. 2006). If pH is higher than 9.0, bicarbonate ions are converted to carbonate ions, which are the scavengers for ^oOH radical that slow down the kinetic rate of the oxidation reaction (Kurniawan et al. 2006). Carbonate ion can destroy ^oOH radical stronger than bicarbonate. In addition, the present of Cl^- also scavenges ^oOH radical. In this study, Anion concentration of Cl^-, HCO_3^- and CO_3^{2-} were high in landfill leachate (table 2). Therefore, at pH higher than 9.0, the oxidation of organic carbon compounds is gone down by scavengers existing in landfill leachate.

$$^{o}OH + P \rightarrow \text{end products} \quad (3)$$

where P represents the scavenger of hydroxyl radical (Kurniawan et al. 2006a). Some examples of the reactions are presented as follows (Gottschalk et al. 2010):

$$^{\circ}OH + Cl^{-} \rightarrow ClOH^{\circ-}, (\text{k} = 4.3 \text{ x } 10^{9} \text{ M}^{-1}\text{s}^{-1}) \quad (4)$$

$$^{\circ}OH + HCO_3^{\ -} \rightarrow HCO_3^{\ \circ} + OH^{-}, (\text{k} = 1.5 \text{ x } 10^{7} \text{ M}^{-1}\text{s}^{-1}) \quad (5)$$

$$^{\circ}OH + CO_3^{\ 2-} \rightarrow CO_3^{\ \circ} + OH^{-}, (\text{k} = 4.2 \text{ x } 10^{8} \text{ M}^{-1}\text{s}^{-1}) \quad (6)$$

The experimental results of this study show that pH 8-9 was optimal for oxidation process by O_3/H_2O_2. These inhibitors often terminate the chain reaction and inhibit ozone decay (Gottschalk et al. 2000). Therefore, in this research, the efficiency of COD and TOC removal decreased at high pH of leachate. The result of this investigation was different to several authors such as Benitez et al., 2000 and Ratanatamsku et al., 2009, as optimum results was attained at pH 11. However, it is similar to some researchers such as Aken et al., 2011, where the pH value of 9 was achieved; Fang et al., 2005, optimal pH was from 7 to 8; Kurniawan et al, 2006b, optimal pH was from 8 to 9.

Effect of Reaction Time

From the preceding experiments on effects of pH of landfill leachate, pH range of 8-9 was the optimum choice for the experiments of reaction time. Due to the initial pH of landfill leachate of about 8.20 on average, the next experiments should adjust and maintain pH to about 8 in order to not waste intense chemicals. The experiments were performed three times (Experiment 1, Experiment 2 and Experiment 3) with a reaction time range of 40-140 min and are shown in the figure 3.

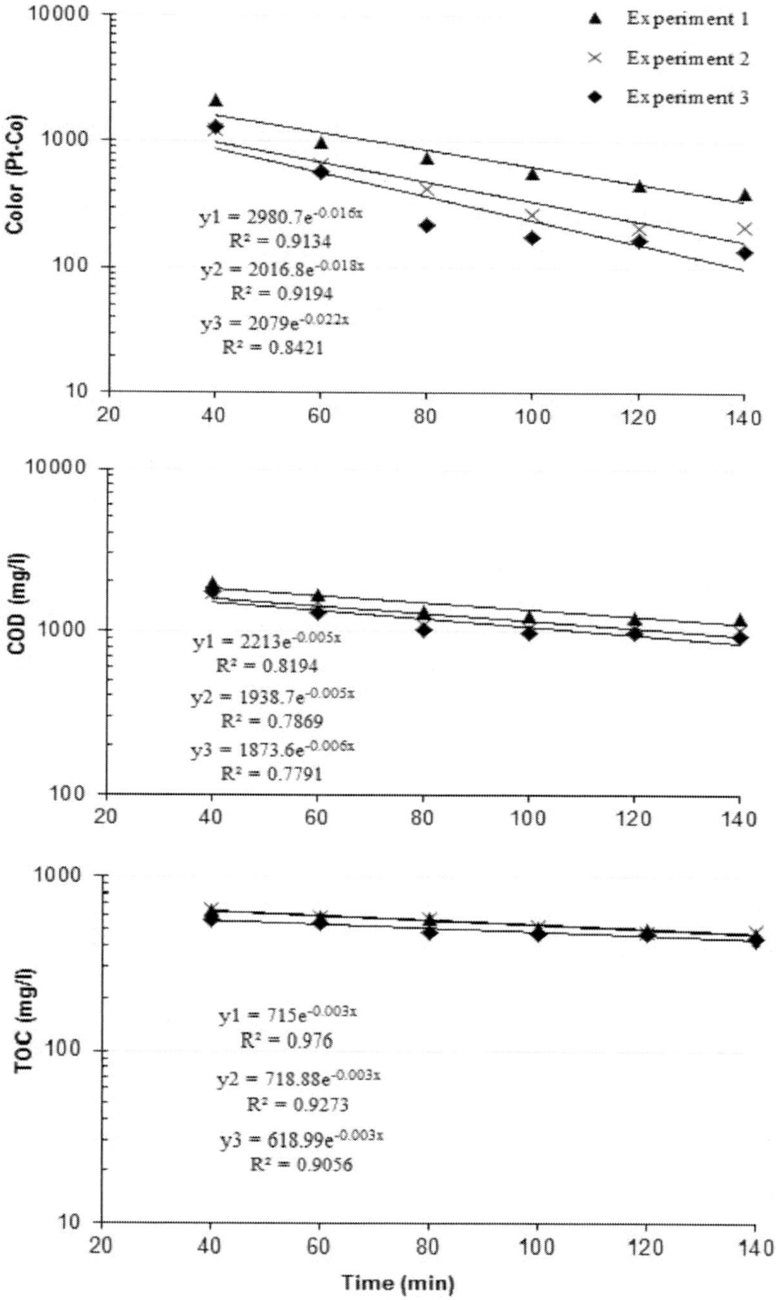

Figure 3. The effect of reaction time on Color, COD and TOC removal from landfill leachate by O_3/H_2O_2 system without CRR at pH 8.

The efficiency was very low in experiments that were conducted with reaction time lower than 40 min.

After coagulation, concentrations of color, COD and TOC were 2,025 – 2,800 Pt-Co, 2,176 – 2,638 mg/l, 725 – 822 mg/l, respectively. The experimental results of landfill leachate treatment by O_3/H_2O_2 showed that when reaction time on landfill leachate increased, the efficiency of COD, color and TOC increased and achieved stable after 80 min.

Treatment result of Color, COD and TOC after flocculation in 80 min of aeration time was 73-89% (reduced to 222 – 748 Pt-Co), 50-53% (reduced to 1,021 – 1,330 mg/l) and 28-34% (reduced to 478 – 579 mg/l), respectively. Performance of color treatment was very high and the efficiency of COD removal was also quite good after 80 min. Continuing to increase the treatment time did not significantly increase the efficiency, as the reaction may reach saturation. So, if the reaction time increased, the efficiency was not high and the reaction consumed a lot of energy, H_2O_2 and O_3 and time. From the preceding experimental results, an aeration duration of 80 min was chosen as the optimal time for these experimental conditions.

The optimal of reaction time in this study was lower than that of the study by Ratanatamskul et al., (2009). It showed that after 180 min of reaction, COD and color removal were 50 and 90%, respectively. The efficiency of the study was higher but it consumed more time for treatment. The result of study from Oliveira at el., (2011) showed that with combination of O_3/H_2O_2, after 60 min and ozone dosage at 5.6 g O_3/h, initial pH 7, and 400 mg/l of hydrogen peroxide, COD removal was 72%. The result of this study indicates that high efficiencies were obtained with less reaction time or ozone dosage.

Effects of CRR

In this study, before treatment by O_3/H_2O_2/CRR, landfill leachate was pre-treated by coagulation with PAC of 1,500 mg/l. The above experiments of pH and reaction time in fluence on the treatment systems of landfill leachate by O_3/H_2O_2 without CRR have determined the optimal pH and reaction time. Therefore, experiments of O_3/H_2O_2/CRR combination were also conducted at the optimal pH and reaction time identified above. In these experiments of specific surface effect of CRR, the conditions were the same with above experiments: airflow generated from Ozone Generator was 7 L/min corresponding ozone concentrations generated about 2.882mg/h and the volume of test sample was 1L. The landfill leachate treatment efficiency was compared with concentrations after flocculation. After coagulation, concentrations of color, COD and TOC were 1,650 – 1,806 Pt-Co, 2,142 – 2,332 mg/l, 955 – 1,010 mg/l, respectively. The experiments were carried out simultaneously on the same concentrations on color, COD and TOC of landfill leachate. The experiments were performed three times, labeled Experiment 1, Experiment 2 and Experiment 3, with following conditions: reactor without CRR and 3 contactors with CRR having specific surface of 356, 539 and 728m^2/m^3, respectively. Experimental results are shown in Figure 4.

The experimental results of effects of CRR specific surface with the conditions as above showed that the removal of original pollutants did not change much between experiments with the reactor systems without CRR and the reactor systems with CRR up to 356 m^2/m^3 of specific surface. There was COD removal (the reactor without CRR: 33-37%, the reactor with CRR 356m^2/m^3: 35-38%), TOC removal (the reactor without CRR: 29-40%, the reactor with CRR 356m^2/m^3: 30-41%) and color removal (the reactor without CRR: 65-75%, the reactor with CRR 356m^2/m^3: 68-80%). However, when the specific surface increased over 539m^2/m^3, the removal of organic compounds in landfill leachate in this study increased significantly. Especially the removal of COD and TOC achieved the highest performance on the reactor with CRR specific surface of 728m^2/m^3. The treating performances for color, COD and TOC reached the highest achievement of 75-86%, 48-52% and 39-48%, respectively, in the experimental systems with CRR. The efficiency was not really high in reactor with CRR up to 356 m^2/m^3, O_3 was an agent poorly soluble in water from aeration in this study.

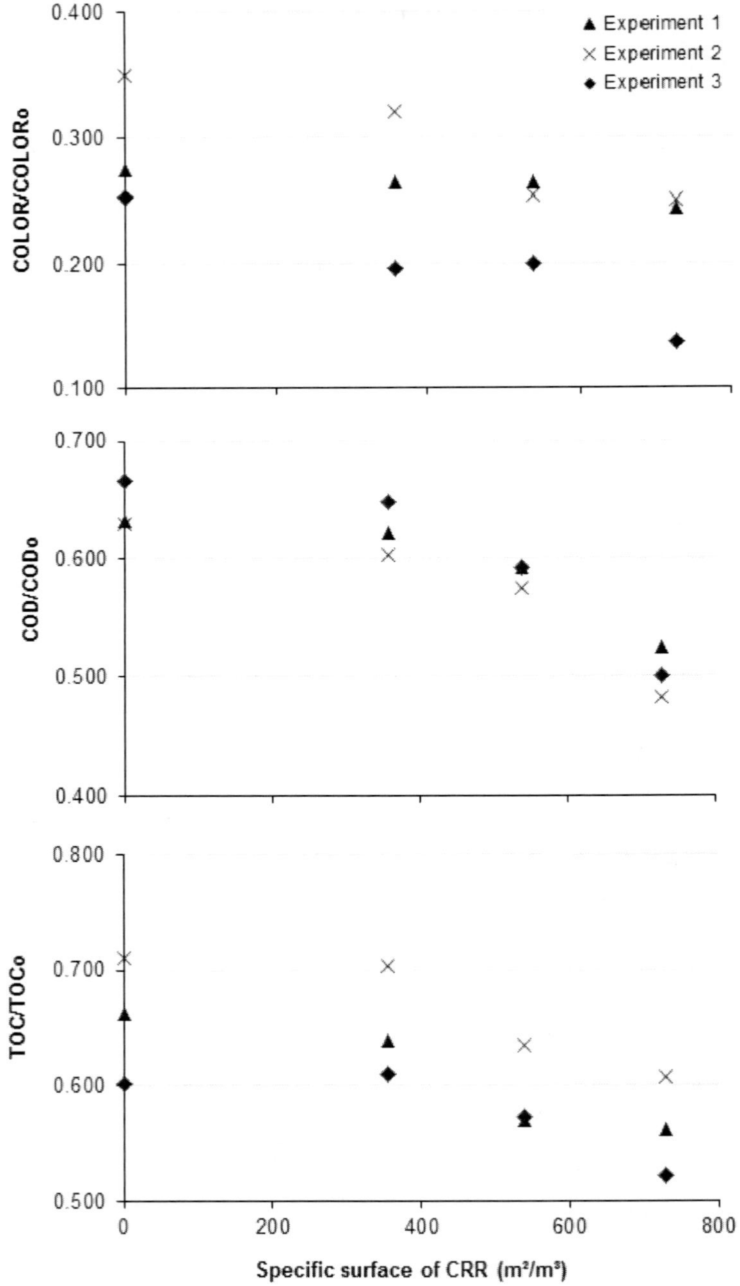

Figure 4. The effect of CRR specific surface on Color, COD and TOC removal from landfill leachate by O_3/H_2O_2/CRR system at pH 8 after 80 min.

It was not so different from the experiments of the reactor without CRR. Only when using the reactor containing CRR with specific surface of 539m^2/m^3, performances increased significantly. In the experiment with CRR of 728 m^2/m^3 specific surface, the treating performances for color, COD and TOC were higher than experiments with other CRR, reaching 8%, 14% and 9% and achieved the highest performance in the experiment of this study. As a result, the larger specific surface increased mass transfer of the gas phase into the liquid phase. The cause could be explained as following: with the large specific surface of CRR, reaction time of the gas phase and liquid phase increased, that resulting in increased O_3 transfer from the gas phase into the liquid phase. Therefore, since the amount of O_3 in landfill leachate was higher, the reaction of O_3 with organic compounds in landfill leachate was also enhanced. This result was substantiated via the results shown in table of analysis of the amount of dissolved O_3 in experiments where O_3 and CRR combined.

Table 3 presents ozone utilization efficiency, in experiments with the CRR reactor, the amount of ozone was used higher than that in the reactor without CRR. The highest ozone utilization efficiency, from 70.93% to 73.21%, was achieved in experiments with CRR of 728 m^2/m^3 specific surface.

This result caused by CRR that extended reaction time for not only O_3 and H_2O_2 but also O_3 and organic compounds from landfill leachate. It means that CRR can enhance mass transfer of ozone molecule from gas phase into liquid phase. At this condition, after 80 min of reaction, the average amount of O_3 generated was 4.803g.

Table 3. Ozone utilization efficiency in experiments of landfilll leachate treatment by combination of O_3/H_2O_2/CRR

Specific surface of CRR (m^2/m^3)	Ozone utilization efficiency (%)		
	Experiment 1	Experiment 2	Experiment 3
0	50.04	48.18	55.44
356	51.92	51.73	58.64
539	57.15	56.90	64.51
728	70.93	72.72	73.21

The average reduction of COD was 1,117 mg/l of landfill leachate, the average amount of O_3 was consumed to treatment for 1kg COD was 3.441 kg O_3/kg COD. CRR are expensive, however, they can be used for a long time. Therefore, they can be considered economically for landfill leachate treatment. Some previous studies showed the level of consumption of O_3 to treat COD landfill leachate such as Tizaoue et al., (2007), and Wang et al., (2003) used 1.5 and 16kg O_3/kg COD when using in combination to $O_3/H_2O_2$to treat landfill leachate, while Ho et al., (1994) used less consumption of 1kg O_3/kg COD. Aziz et al., (2013) showed 1.6 kgO_3/kg COD of consumption when using in combination to O_3/Fenton to treat landfill leachate.

Conclusion

In this study, the experimental results identified the optimal pH range of 8-9 for landfill leachate treatment by O_3/H_2O_2. Optimal reaction time was 80 min. Specially, the experiments of combination O_3/H_2O_2/CRR had significantly increase in the treatment performance of organic compounds in landfill leachate with CRR specific surface of 728 m^2/m^3. The removals of color, COD and TOC were over 10% higher than that in the experimental systems without CRR. O_3 solubility from aeration into the liquid phase of experiments with CRR of 728 m^2/m^3 of surface was also much higher than that in experiments without CRR. The combination O_3/H_2O_2/CRR was a treating solution that had good results in treating landfill leachate. Experiments also showed that 3.441 kg O_3 was required to treat 1 kg COD.

References

Aken, V.P., Lambert, N., Degreve, J., Lier, S., Luyeten, J., 2011. Comparison of different oxidation methods for recalcitrance removal of landfill leachate. *Ozone: Science and Engineering,* 33: 294-300.

Alwasit, A.A., Riyadh S. Al-mukhtar and Omar J. Mahmood, 2009. Mass transfer of ozone in purified water". *Eng. And Tech. Journal*, 27(4): 776-786.

American Public Health Association (APHA), Standard Methods for the Examination of Water and Wastewater. 19th American Public Health Association, Washington DC. 2012.

Aziz, H.A., Salem, S., Abu. A.,2012. Performance of Ozone/Fenton in the Advanced Oxidation Process of Semi-Aerobic Landfill Leachate. *The Asian Conference on Sustainability, Energy and the Environment Official Conference Proceeding*. 202- 212.

Benitez, F.J., Beltran-Heredia, J.B., Acero, J.L., Rubio, F.J., 2000. Contribution of free radicals to chlorophenols decomposition by several advanced oxidation processes. *Chemosphere*, 41: 1271 - 1277.

Cataldo, F., Angelini, G. 2013. Municipal landfill leachate treatment using the combined action of activated carbon and ozone". *Ozone: Science and Engineering*, 35: 55 - 62.

Fang, H.H.P., Lau, I.W.C. and Wang, P., 2005.Anaerobic treatment of Hong Kong leachate followed by chemical oxidation. *Water Science and Technology*, 52: 41 - 49.

Gottschalk, C., Libra, J.A. and Saupe, A., Ozonation of Water and Wastewater, A practical guide to understanding ozone and its application. WILEY-VCH verlag GmbH, D-69469 Weinheim, Federal republic of Germany. 2010.

Haapea, P., Korhonen, S., Tuhkanen, T., 2002. Treatment of industrial landfill leachates by chemical and biological methods: ozonation, ozonation + hydrogen peroxide, hydrogen peroxide and biological post-treatment for ozonated water. *Ozone: Science and Engineering*, 24: 369-378.

Ho, S., Boyle, W.C., Ham, R.K., 1974. Chemical treatment of leachate from sanitary landfill. *Water Pollution Control Federation*, 46:1776-1791.

Khanh, N.H., Landfill environment and techlonogy of landfill leachate treatment, Publish house of technological Science, in Vietnamese. 2009.

Kurniawan, T.A., Wai-hung Lo, W.H., Chan, G.Y.S., 2006a. Radicals-catalyzed oxidation reactions for degradation of recalcitrant compounds from landfill leachate. *Chemical Engineering Journal*, 125: 35 - 57.

Kurniawan, T.A., W.-H. Lo, and G.Y.S. Chan. 2006b. Degradation of recalcitrant compounds from stabilized landfill leachate using a combination of Ozone-GAC adsorption treatment." *Journal of Hazardous Materials,*B137: 443 - 455.

Li, Q., Zhou, S., Hua, T., 2010. Removal of organic matter from landfill leachate by advanced oxidation processes: A review. *International J. of Chemical Eng*., 1-10.

Ministry of Science and Technology, Viet Nam. 2000. *TCVN 6634:2000: Water quality. Guide to determine total organic carbon (TOC) and soluble organic carbon (DOC).*

Oliveira, R., S. Cortez, P. Teixeira, and M. Mota. 2010. Ozonation as polishing treatment of mature landfill leachate. *Journal of Hazardous Materials,*182: 730 - 734.

Ratanatamskul, C. and Parinya, A., 2009. Removal of COD and colour from old-landfill leachate by Advanced Oxidation Processes. *Int. J. Environment and Waste Management*, 4(3/4): 470-481.

Schulte, P., Bayer, A., Kuhn, F., Luy, Th., and Volkmer, M., 1995. H_2O_2/O_3, H_2O_2/UV and H_2O_2/Fe^{2+} processes for the oxidation of hazardous wastes. *Ozone: Science and Engineering*, 17: 119-134.

Sumanaweera, S., 2004. Advanced oxidation combined with a membrane bio-reactor for landfill leachate treatment. *A master thesis, Asian Institute of Technology, School of Environment*, Resources and Development, Thailand.

Tizaoui, C., Bouselmi, L., Mansouri, L., Ghrabi, A., 2007. Landfill leachate treatment with ozone and ozone/hydrogen peroxide systems. *Journal of Hazadous Material,*140:316-324.

Wang, F., Smith, D.W., El-Din, M.G., 2004. Oxidation of agen raw landfill leachate with O_3 only and O_3/H_2O_2: treatment efficiency and molecular size distribution analysis. *Science and Engineering,* 26: 287-298.

Wu, J.J, Chih-Chao, W., Hong-Wen, M., Chia-Chi C., 2004. Treatment of landfill leachate by ozone-based advanced oxidation processes. *Chemosphere,* 54: 997 - 1003.

Zhou, S., Huo-sheng Li, Yan-bo Sun, Ping Feng, Jing-da Li., 2009. Advanced treatment of landfill leachate by a new combination process in a full-scale plant. *Journal of Hazardous Materials*, 172: 408 - 415.

In: Leachates
Editor: Maria R. Mattox
ISBN: 978-1-53612-508-5

Chapter 3

PHYTOREMEDIATION OF MUNICIPAL LANDFILL LEACHATE IN CONSTRUCTED WETLAND: TOXICOLOGICAL CHARACTERISTICS AND POTENTIAL IMPACTS ON THE ENVIRONMENT

Hamidi Abdul Aziz[1,*] and Hossein Farraji[1, 3]
[1]School of Civil Engineering, Engineering Campus, Universiti Sains Malaysia, Nibong Tebal, Penang, Malaysia
[2]Solid Waste Management Cluster, Engineering Campus, Universiti Sains Malaysia, Penang, Malaysia
[3]Institute of Standard and Industrial Researches of Iran (ISIRI) Mashhad, Khorasan Razawi, Iran

[*] Corresponding Author Email: Cehamidi@usm.my.

ABSTRACT

Landfill leachate (LL) is a high-strength wastewater and poses a potential environmental threat if not managed properly. High concentrations of hazardous heavy metals, phenols, and ammonia nitrogen are concerns in LL. Plant application in aquatic media as a sustainable and eco-friendly technique is a potential biological treatment method of LL. This chapter introduces the main hazardous compounds of LL and their expected effects on the environment. Plant-based treatment methods (phytoremediation and constructed wetland) for LL are also discussed.

Keywords: ammonia, adsorbent, COD, metallic element, color, phenolic compound

1. PHYTOREMEDFIATION AND CONSTRUCTED WETLAND

Treatment of landfill leachate (LL) for solid waste management is costly (Lavagnolo et al., 2016). Sustainable waste treatment as a suitable waste disposal (Farraji et al., 2015a) and wastewater treatment by phytoremediation technique as an emerging green technology are the major green technologies in the world. The overpopulation in Asia and Africa has emphasized the need for cost-effective LL treatment methods for environmental protection (Farraji et al., 2016a). Phytoremediation is commonly used for wastewater treatment in Europe, and constructed wetland (CW) is the most cost-effective high-performance treatment method for decontamination of LL (Stefanakis et al., 2014).

Application of free water surface for LL treatment is common in North America and Europe (Vymazal, 2010). Two types of CW as promising biologically degradation treatment methods (phytoremediation) for LL are subsurface flow CW and surface flow CW. Subsurface flow with wastewater flow through the bed of CW is preferred for managing the odor problem. LL is a worldwide environmental concern (Renou et al., 2008) and can be solved by cost-effective and eco-friendly treatment methods, such as CWs (aquatic phytoremediation).

Phytoremediation is an emerging green and environmental-friendly treatment technique and is widely used for different types of wastewaters, such as domestic wastewater (Thongtha et al., 2014; Fountoulakis et al., 2017), dye (Khandare & Govindwar, 2015; Rane et al., 2015), mine (Mishra and Shukla, 2016; Türker et al., 2014), palm oil mill effluent (Darajeh et al., 2014), olive oil mill effluent (Abdallah & Quot, 2016), pharmaceuticals (Carvalho et al., 2014), livestock (Almeida et al., 2016), and LL (Madera-Parra et al., 2014; Madera-Parra et al., 2015). CW is an effective treatment method for wastewater in aquatic-plant-based decontamination (Gorito et al., 2017). The application of this environmental-friendly technique depends on the concentration of contaminants, such as suspended solids (Aiello et al., 2016) which can clog the pores of CW media for root growing, and toxic compounds (Fujii et al., 2017) which can prohibit plant growth and development. Transforming of pollutants to less toxic ones, mineralization, or uptaking by plant tissues is an urgent requirement for wastewater treatments (Gorito et al., 2017). Given the low concentration of contaminants and suspended solids or absence of toxic compounds in domestic wastewater, phytoremediation is commonly used to treat municipal wastewater (Zhang et al., 2014; Zhang et al., 2015; Istenic et al., 2015). Meanwhile, LL contains a high concentration of hazardous metallic elements and toxic organic and inorganic compounds, such as ammonia nitrogen and phenols (Renou et al., 2008). This chapter discusses pollutant removal from LL by CW through phytoremediation. Each of the pollutants is described, and the toxicity of these pollutants in the environment is discussed.

2. COD and Color Removal

The highest concentration of chemical oxygen demand (COD) is caused by food wastes (Bhatt et al., 2016). Same as ammonia nitrogen, only the specific chain of combined physicochemical treatment methods, such as ultrasonication, Fenton oxidation, and ozonation process, can present high decontamination for COD (95%) and color (100%) in a lab-scale study (Asaithambi et al., 2017). The said high-cost methods are applied in industrial scale because of their high removal efficiency even with pre-treatment by activated carbon (Poblete et al., 2017). The single part of Fenton oxidation cannot achieve effective removal for COD (47%) and color (74%) but is suitable for acidic effluent (pH = 2.5) unlike costly treatment processes (Kocakaplan et al., 2016). Natural and synthetic adsorbents are unsuitable for LL decolonization (Farraji et al., 2015b; Mojiri et al., 2014). Vertical flow CW planted with *Typha domingenosis* obtains low-cost effective removal of color (90.3%) and COD (86.7%) (Mojiri et al., 2016), and horizontal flow CW through LL decontamination achieves excellent performance (Garbo et al., 2017). Current pilot-scale studies demonstrated that practical and performance aspect (60–90% removal for all characteristics) of CW utilization for LL treatment (Madera-Parra et al., 2015) and plant distribution as well as suitability of local species play the main roles in the CW removal performance.

3. Nutrient Removal

Zhang et al. (2007) indicated that plant uptaking is the main mechanism for removing nutrients in CWs. The design of treatment system exerts a critical effect on the optimization of plant density (McCutcheon & Schnoor, 2003). Iamchaturapatr et al. (2007) mentioned that the planted CW and the plant geometries (such as roots and shoots) should be appropriately selected for high-efficiency design in aquatic media. Biogeochemical parameters (oxidation–reduction potential and dissolved oxygen) are auspicious for nitrogen removal procedures, such as macrophyte plant uptake, ammonification, and nitrification/denitrification (Koottatep & Polprasert 1997). Farid et al. (2014) indicated that polycultures (species mixtures) can achieve better CW nutrient removal than monocultures. Tanner et al. (1995) detected 20% nitrogen and 60% phosphorus removal in CW planted with *Schoenoplectus validus* compared with unplanted CW. Lin et al. (2002b) proposed CW systems planted with

Commelina communis, *Ipomoea aquatica*, *Pennisetum purpureum*, *Pistia stratiotes*, and *Phragmites australis*. They reported that 89%–96% of nitrogen are removed due to denitrification and 4%–11% of nitrogen are removed due to vegetation uptake. The plant effects on denitrification depend on the ability of plants to produce carbon for denitrification (Lin et al., 2002a). Bachand and Horne (1999) used a mixture of submerging or floating and a large amount of emergent plants or grasses to improve denitrification rates in organic carbon-limited free-surface wetlands. Langergraber (2005) reported that nutrient uptake by macrophytes exhibit no major effect on water treatment. Taylor et al. (2011) found that COD removal decreases in unplanted CW whereas does not change in planted CW during cold season. Ouellet-Plamondon et al. (2006) obtained slight COD removal in summer, and Garfí et al. (2012) observed high biological oxygen demand (BOD) removal in summer. The ability or interest in plant species for nutrient removal in CWs differs; for example, *C. indica* exhibits significant affinity to nitrate, and *S. validus* favors NH_4-N (Zhang et al., 2009). In summary, macrophytes play three significant roles in simulated wetlands (Brisson & Chazarenc, 2009):

- Uptake nutrients and adsorb and accumulate metallic elements and poisonous materials from wastewater
- Transfer oxygen to roots (rhizosphere) to facilitate the normal growth, propagation, and decomposition of microorganisms
- Increase and stabilize the hydraulic transference of media.

3.1. Ammonia Nitrogen

High concentration and toxicity effect of ammonia are influencing factors of biological and microbiological treatments (Lin et al., 2007; Tan et al., 2016). Ammonia stripping is a pre-treatment process for LL (Cheung et al., 1997). Finding tolerant organisms, microalgae, and macrophytes to ammonia is the main aim of phytoremediation (Budi et al., 2016; Svensson et al., 2005). Through physicochemical treatment, only high-performance aerobic sequencing batch reactors augmented with synthetic adsorbents, such as ZELIAC (which is a combination of zeolite, activated carbon, limestone, rice husk ash, and Portland cement), can effectively remove ammonia nitrogen (97.97%) (Mojiri et al., 2014). Vertical CW presents high efficiency in ammonia removal, while horizontal CW achieves high performance in COD removal (Yalcuk & Ugurlu, 2009). Macrophytes use ammonia nitrogen as a macro element (nutrient); thus, the availability of phosphorus as other critical growing element is limited through LL phytoremediation in CW (Paskuliakova et al., 2016). Artificial aeration can enhance the 11%–99% removal efficiency of ammonia nitrogen, and aerated wetland is a cost-effective treatment system for ammonia removal (Butterworth et al., 2013) and can produce high nitrogen removal at high temperature (>14.2°C) (Mietto et al., 2015).

3.1.1. Toxicity of Ammonia

Ammonia nitrogen (NH_4-N) is a major nitrogen-containing hazardous compound and is present in aquatic and terrestrial media (Kim et al., 2016). The presence of ammonia in aquatic media is the main cause for Macrobrachim and Macroinvertebrate growth (Gomes et al., 2016; Willingham et al., 2016). The fertilized eggs and larvae of these species are sensitive to ammonia toxicity (El-Greisy et al. 2016). Meanwhile, Chen et al. (2016) reported the resistance of shrimps to ammonia concentration. Ammonia is the main cause for plant toxicity in LL phytoremediation (Paskuliakova et al., 2016). Consequently, a suitable, effective, and green decontamination project highly depends on the pre-treatment of ammonia (Poveda et al., 2016), especially in high concentration of this toxic nitrogen compound.

3.2. Phosphorus

Phosphorus is a compound finite in natural rocks and is widely applied in industry and agriculture (Jordaan et al., 2010). Phosphorus is also found in wastewater released to the environment and exerts numerous environmental impacts (Liu et al., 2013). The recovery of phosphorus is important to save the natural sources of phosphorus and the environment (Zhou et al., 2015). Phosphorus removal can be achieved by sedimentation as the main physical decontamination, adsorption as the major chemical method, and biological decontamination (Yalcuk & Ugurlu, 2009). Direct uptaking by macrophytes (cattail) (Bonomo et al., 1997), uptaking by plant and microorganisms (Sakadevan & Bavor, 1998), and uptaking by microalgae (Pereira et al., 2016) are other phosphorus removal methods. Notably, phosphorus removal is less sensitive to temperature conditions (Wittgren & Mhlum, 1997). A positive correlation exists between phosphorus removal and treatment time (Rousseau et al., 2004), and low concentration of phosphorus contributes to biomass production and LL decontamination performance (Yalcuk & Ugurlu, 2009). The electro-Fenton method combined with sequencing batch reactor provides high efficiency in phosphorus removal and enhances the biodegradability (BOD/COD) of LL (Lin & Chang, 2000). Given that decontaminating carbon sources is the main aim of LL, microbial communities with high ability to uptake and high concentration of carbon sources exert a strong influence on performance. Phosphorus is not release in these microbial communities because of the absorption and storage process of carbon sources; thus, these microbial communities compete against microorganisms which accumulate biologically stored phosphorus (Winkler et al., 2011). High biodegradable proteins of COD in LL play a major role in improving biological phosphorus removal in a combined 2.5% municipal wastewater (Yuan et al., 2016).

4. PHENOLS

Phenols are highly toxic organic compounds and exert hazardous effects on aquatic bio-systems even in low concentrations; they also easily degrade in high pH values (Vázquez et al., 2006). Aziz et al. (2010) found phenolic compounds in LL. High persistence of these compounds causes high environmental impacts (Boonnorat et al., 2014; Kalmykova et al., 2013). Studies on phenol removal by CW are rarely cited and are limited. Fujii et al. (2017) considered phenol removal as target decontamination for LL in CW. The efficiency of phenol removal in CW can be accepted as capable performance for even strength treatment system. CW obtains 83.4% removal for olive oil wastewater (Bubba et al., 2004) and 90.5% for LL (Mojiri & Ziyang, 2015). Compared with electromagnetic reactor (47.1% degradation) (Yuan et al., 2016), ozone (80.17% removal) (Wang et al., 2016), and electro-microbial membrane (99.6% removal) (Cao et al., 2017) applied for high concentration phenol wastewater, CW is a cost-effective, green, and environmental-friendly treatment method. Works on phenol removal from wastewater are fully reviewed by Villegas et al. (2016).

4.1. Toxicity of Phenols

The US Environmental Protection Agency has set a standard strength (<1 ppb) for phenols in surface water (Kazemi et al., 2014). Their toxicity level in humans and aquatic media should be 9–25 mg/L (Kulkarni & Kaware 2013). They can be found in compounds such as chlorophenol, bisphenol A, and phenolic endocrine-disrupting compounds (Zhang, 2014). Abnormal development and growth reduction in animal tissues of the nervous system, kidney, liver, and heart are related to phenol toxicity (Mukherjee et al., 2013).

5. Heavy Metals

Heavy metal removal from wastewater in CW with the assistance of plant species has been widely investigated. Some studies (Lim et al., 2003) reported no or insufficient effect of plants in metallic element removal, whereas other studies (Lim et al., 2003) reported extraordinary effects of plants in removing Zn, Pb, Cd, and Cu. Zayed et al. (1998) reported that duckweed shows potential for removal of Cd, Se, and Cu from contaminated wastewater. Maine et al. (2009) used a CW system planted with *Typha domingensi* for high pH and conductivity wastewater containing metals (Cr, Ni, and Zn), and showed that the treatment process removes *Echhornia crasipes*. Therefore, plant selection is important for CW systems. The influencing factors of metal accumulation by wetland macrophytes are metal concentration, pH, and nutrient position in the substrate (Deng et al., 2004). Acid mine drainage containing Mn, Zn, Cu, Ni, B, and Cr is collected in natural and CW plants; nonetheless, only a small percentage of heavy metals are accumulated in plant bodies (Mays & Edwards 2001). The CW removal efficiencies for lead, cadmium, iron, nickel, chromium, and copper are from 40.9% to 91.9% (Khan et al., 2009). Mufarrege et al. (2015) studied CWs with plants of high concentration of Cr, Ni, and Zn, and their results showed that a negative correlation exists between nitrogen and phosphorus amount and metallic element accumulation. They also found that metals are retained and adapted after accidental dumping of high concentration of heavy metal plants. According to Phillips et al. (2015), plant species can be used as metal indicators. Therefore, plant ability in CW should be further studied.

The main performance and eco-friendly aspect of phytoremediation for LL treatment is attributed mainly to the non-degradable contaminants of LL. Metallic elements of major hazardous effects on human food chain and living organisms are the most critical toxic elements of LL (Oka et al., 2017), and macrophytes are the heavy metal accumulators (hyperaccumulators) (Madera-Parra et al., 2014). Previous holistic reviews (Baun & Christensen, 2004; Öman & Junestedt, 2008; Renou et al., 2008) indicated that the concentrations (mg/L) of metallic elements in LL are as follows: Fe = 2100, Cd = 0.13, Zn = 120, Mn = 65, Ni = 3.2, Cr = 1.6, and Pb = 1.5; these concentrations gradually decrease with the aging of the LL (Kulikowska & Klimiuk, 2008; Kjeldsen et al., 2002). For more than two decades, metallic element removal from LL have been extensively investigated in a lab scale (Lavrova & Koumanova, 2010; Sun & Austin,

2007; Mojiri & Ziyang, 2015; Mojiri et al., 2016), pilot scale (Madera-Parra et al., 2015), and full scale (Mietto et al., 2015; Aiello et al., 2016).

5.1. Toxicity of Metallic Elements

Metallic elements (heavy metals) are recognized as a group of elements with 63.5–200.6 atomic weight and a specific gravity higher than 5 (Lakherwal, 2014). The toxicity, health effect, and mechanism of numerous heavy metals have been previously reviewed (Jaishankar et al., 2014). Impairing effect on reproduction and respiration of the coral reef (Thompson, 2016), coastal sediment pollution (Polidoro et al., 017), and bacterial resistance to antibiotic in aquatic media (Zhou et al., 2015) are some of the environmental impacts caused by the toxicity of metallic elements. Recently, Baqar et al. (2016) illustrated bioaccessibility of heavy metals on soil, ground water, and edible leafy vegetables in Lahore. Their results indicated that bioaccumulation of metallic elements can be a critical human health problem in the near future.

6. Removal Mechanisms

Water penetration from wastes in landfills creates leachate containing ammonia nitrogen, metallic elements, COD, BOD, color, odor, dioxins, phenolic compounds, and suspended solids (Taulis, 2005). Physicochemical and biological treatment processes cannot meet acceptable water quality standards (Torretta et al., 2016). CW is a cost-effective and eco-friendly treatment method especially in tropical areas where phytoremediation presents high annual efficiency for decontamination (Akinbile et al., 2012; Yuka Ogata et al., 2015; Sim et al., 2013). High performance of CW among organic pollutants (such as ammonia nitrogen), COD, and color is the advantage of this emerging decontamination method (Farraji et al., 2016b).

A complex decontamination process is a suitable mechanism for pollutant removal in LL and highly depends on filtration, sedimentation, volatilization, biodegradation in various microbial process, adsorption on capable adsorbent, precipitation, and plant uptake (Vymazal, 2007; Kadlec & Wallace, 2009; Faulwetter et al., 2009; Fujii et al., 2017). Pollutant removal in CW can be achieved by biological, chemical, and physical methods (Figure 1). The specific mechanism of heavy metal removal from LL in CW is presented in Figure 2. The aforementioned mechanisms are influenced by environmental conditions during decontamination (Vymazal & Kröpfelová, 2009; Garcia et al., 2010; Huang et al., 2013).

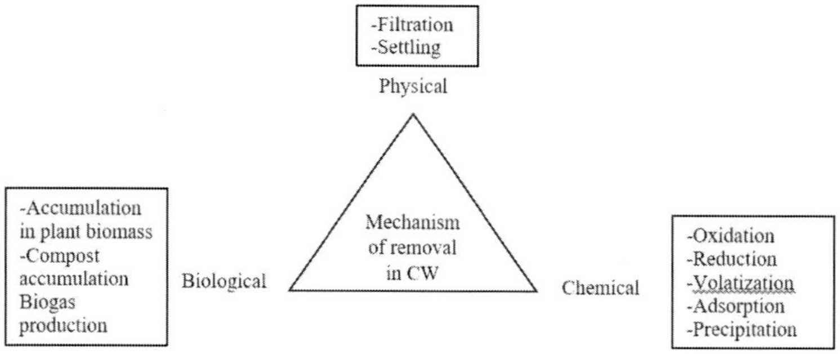

Figure 1. Pollutant removal mechanisms in CW (Vymazal, 2007; Kadlec & Wallace, 2009; Faulwetter et al., 2009; Fujii et al., 2017).

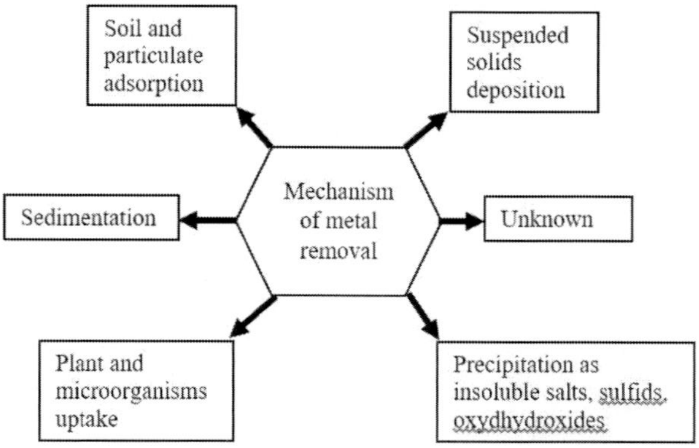

Figure 2. Mechanisms of metallic element removal in CW (Akinbile et al., 2012; Batty et al., 2002; Cheng et al., 2002; Oka et al., 2017; Marchand et al., 2010).

7. Plant Density

Plants as the biological degradation, storage organization, translocation, and removal member of CW exert a major effect on the decontamination process. Consequently, the density of macrophyte application in CW exhibits a significant influence on the removal efficiency (Madera-Parra et al., 2015). This issue in LL treatment in CW needs further exploration. Table 1 illustrates numerous effects of plant density on CW for other types of wastewaters.

Table 1. Plant density in CW literature

Effect	References
Low-density CW achieves high phosphorus removal	(Lin et al., 2002)
Ammonia removal is insensitive to plant density	(Yousefi & Mohseni 2010)
Mixed cultures supply higher plant density in CW than monocultures	(Liang et al., 2011)
Low density of *Vallisneria spiralis* is suitable for PAH[1] degradation	(Liu et al., 2014)
High plant density causes low growth rate; thus, surface area per plant is a key plant density parameter for PAH[1] degradation	(Meng & Chi, 2015)
Pteris vittata has no significant response to arsenic depletion in different densities	(Tu et al., 2004)
Optimum density of *Alyssum murale* for nickel phytomining is 4 plant/m^2	(Bani et al., 2015)
Plant density is highly correlated with Cd removal by Salix	(Greger & Landberg, 2015)

[1] Polycyclic aromatic hydrocarbons.

Conclusion

In this chapter, numerous toxic compounds and elements introduced by LL and its hazardous environmental effect are discussed. Furthermore, the ability of phytoremediation as a green, eco-friendly, and biological treatment for LL treatment is assessed. Current literature indicates that future treatment methods especially in developing countries will be highly concentrated on environmental-friendly methods for municipal LL because of their cost effectiveness and high efficiency as well as easy application and independence to high technological requirements. Suitable environmental conditions especially tropical weather will promote the influence and attraction of plant-based treatment methods for LL. Enhancing plant resistance to high concentration and toxicity of LL and combining phytoremediation with other traditional pre-treatment methods are possible directions for future research.

ACKNOWLEDGMENTS

This work is funded by Universiti Sains Malaysia under Iconic Grant Scheme (Grant no. 1001/CKT/870023) for research associated with the Solid Waste Management Cluster, Engineering Campus, University Sains Malaysia. The author appreciates the Ph.D. research opportunity provided by the Institute of Standards and Industrial Researches of Iran.

REFERENCES

Abdallah, "M. A. R. Y."; Soliman (2016) Phytoremediation of organics and metals from Olive Mill Wastewater (*Doctoral dissertation*).

Aiello, R., Bagarello, V., Barbagallo, S., Iovino, M., Marzo, A., and Toscano, A. (2016) Evaluation of clogging in full-scale subsurface flow constructed wetlands. *Ecological Engineering*, 95, 505–513.

Akinbile, C. O., Yusoff, M. S., and Ahmad Zuki, A. Z. (2012) Landfill leachate treatment using sub-surface flow constructed wetland by Cyperus haspan. *Waste Management,* 32(7), 1387–1393.

Almeida, C. M. R., Carvalho, P. N., Fernandes, J. P., Basto, M. C. P., and Mucha, A. P. (2016) "Constructed Wetlands for Livestock Wastewater Treatment: Antibiotics Removal and Effects on CWs Performance" in Phytoremediation. Cham, *Springer* International Publishing, 267–281.

Bani, A., Echevarria, G., Zhang, X., and Benizri, E. (2015) The effect of plant density in nickel-phytomining field experiments with Alyssum murale in Albania. *Australian Journal of Botany*, 8, 72–77.

Batty, L. C., Baker, A. J. M., and Wheeler, B. D. (2002) Aluminium and Phosphate Uptake by Phragmites australis: the Role of Fe, Mn and Al Root Plaques. *Annals of Botany*, 89(4), 443–449.

Carvalho, P. N., Basto, M. C. P., Almeida, C. M. R., and Brix, H. (2014) A review of plant-pharmaceutical interactions: from uptake and effects in crop plants to phytoremediation in constructed wetlands. *Environmental Science and Pollution Researc*h. *21*(20), 11729-11763.

Cheng, S., Grosse, W., Karrenbrock, F., and Thoennessen, M. (2002) Efficiency of constructed wetlands in decontamination of water polluted by heavy metals. *Ecological Engineering*, 18(3), 317–325.

Darajeh, N., Idris, A., Truong, P., Abdul Aziz, A., Abu Bakar, R., and Che Man, H. (2014) Phytoremediation Potential of Vetiver System Technology for Improving the Quality of Palm Oil Mill Effluent. *Advances in Materials Science and Engineering*, 2014, 1–10.

Faulwetter, J. L., Gagnon, V., Sundberg, C., Chazarenc, F., Burr, M. D., Brisson, J., Camper, A. K., and Stein, O. R. (2009) Microbial processes influencing performance of treatment wetlands: A review. *Ecological Engineering*, 35(6), 987–1004.

Fountoulakis, M. S., Sabathianakis, G., Kritsotakis, I., Kabourakis, E. M., and Manios, T. (2017) Halophytes as vertical-flow constructed wetland vegetation for domestic wastewater treatment. *Science of The Total Environment*, 583, 432–439.

Fujii, D., Soda, S., Machimura, T., and Ike, M. (2017) Removal of phenol, bisphenol A, and 4-tert-butylphenol from synthetic landfill leachate by vertical flow constructed wetlands. *Science of The Total Environment*, 578, 566–576.

Gorito, A. M., Ribeiro, A. R., Almeida, C. M. R., and Silva, A. M. T. (2017) A review on the application of constructed wetlands for the removal of priority substances and contaminants of emerging concern listed in recently launched EU legislation. *Environmental Pollution*, 227, 428–443.

Greger, M. and Landberg, T. (2015) Novel Field Data on Phytoextraction: Pre-Cultivation With *Salix* Reduces Cadmium in Wheat Grains. *International Journal of Phytoremediation*, 17(10), 917–924.

Istenic, D., Bodík, I., and Bulc, T. (2015) Status of decentralised wastewater treatment systems and barriers for implementation of nature-based systems in central and eastern Europe. *Environmental Science and Pollution Research*, 22(17), 12879–12884.

Kadlec, R. H. and Wallace, S. (2009) *Treatment Wetlands*,

Khandare, R. V. and Govindwar, S. P. (2015) Phytoremediation of textile dyes and effluents: Current scenario and future prospects. *Biotechnology Advances*, 33(8), 1697–1714.

Lavagnolo, M. C., Malagoli, M., Garbo, F., and Pivato, A. (2016) Lab-scale phytotreatment of old landfill leachate using different energy crops. *Waste Management*, 55, 265–275.

Liang, M.-Q., Zhang, C.-F., Peng, C.-L., Lai, Z.-L., Chen, D.-F., and Chen, Z.-H. (2011) Plant growth, community structure, and nutrient removal in monoculture and mixed constructed wetlands. *Ecological Engineering*, 37(2), 309–316.

Lin, Y.-F., Jing, S.-R., Lee, D.-Y., and Wang, T.-W. (2002) Nutrient removal from aquaculture wastewater using a constructed wetlands system. *Aquaculture*, 209(1), 169–184.

Liu, H., Meng, F., Tong, Y., and Chi, J. (2014) Effect of plant density on phytoremediation of polycyclic aromatic hydrocarbons contaminated sediments with Vallisneria spiralis. *Ecological Engineering*, 73, 380–385.

Madera-Parra, C. A., Peña-Salamanca, E. J., Peña, M. R., Rousseau, D. P. L., and Lens, P. N. L. (2015) Phytoremediation of Landfill Leachate with Colocasia esculenta, Gynerum sagittatum and Heliconia psittacorum in Constructed Wetlands. *International Journal of Phytoremediation*, 17(1), 16–24.

Madera-Parra, C. A., Pena, M. R., Pena, E. J., and Lens, P. N. L. (2014) Cr(VI) and COD removal from landfill leachate by polyculture constructed wetland at a pilot scale. *Environmental Science and Pollution Research*, 22(17), 12804–12815.

Marchand, L., Mench, M., Jacob, D. L., and Otte, M. L. (2010) Metal and metalloid removal in constructed wetlands, with emphasis on the importance of plants and standardized measurements: A review. *Environmental Pollution*, 158(12), 3447–3461.

Meng, F. and Chi, J. (2015) Phytoremediation of PAH-contaminated sediments by Potamogeton crispus L. with four plant densities. *Transactions of Tianjin University*, 21(5), 440–445.

Mishra, V. K. and Shukla, R. (2016) "Aquatic Macrophytes for the Removal of Heavy Metals from Coal Mining Effluent" in Phytoremediation. Cham, *Springer* International Publishing, 143–156.

Oka, M., Fujii, Y., Soda, S., Ishigaki, T., Machimura, T., and Ike, M. (2017) Removal of heavy metals from synthetic landfill leachate in lab-scale vertical flow constructed wetlands. *Science of The Total Environment*, 584, 742–750.

Rane, N. R., Chandanshive, V. V., Watharkar, A. D., Khandare, R. V., Patil, T. S., Pawar, P. K., and Govindwar, S. P. (2015) Phytoremediation of sulfonated Remazol Red dye and textile effluents by Alternanthera philoxeroides: An anatomical, enzymatic and pilot scale study. *Water Research*, 83, 271–281.

Renou, S., Givaudan, J. G., Poulain, S., Dirassouyan, F., and Moulin, P. (2008) Landfill leachate treatment: Review and opportunity. *Journal of Hazardous Materials*, 150(3), 468–493.

Thongtha, S., Teamkao, P., Boonapatcharoen, N., Tripetchkul, S., Techkarnjararuk, S., and Thiravetyan, P. (2014) Phosphorus removal from domestic wastewater by Nelumbo nucifera Gaertn. and Cyperus alternifolius L. *Journal of Environmental Management*, 137, 54–60.

Tu, S., Ma, L. Q., Fayiga, A. O., and Zillioux, E. J. (2004) Phytoremediation of Arsenic-Contaminated Groundwater by the Arsenic Hyperaccumulating Fern *Pteris vittata* L. *International Journal of Phytoremediation*, 6(1), 35–47.

Türker, O. C., Türe, C., Böcük, H., and Yakar, A. (2014) Constructed Wetlands as Green Tools for Management of Boron Mine Wastewater. *International Journal of Phytoremediation*, 16(6), 537–553.

Vymazal, J. (2007) Removal of nutrients in various types of constructed wetlands. *Science of The Total Environment*, 380(1), 48–65.

Yousefi, Z. and Mohseni-Bandpei, A. (2010) Nitrogen and phosphorus removal from wastewater by subsurface wetlands planted with Iris pseudacorus. *Ecological Engineering*, 36(6), 777–782.

Zhang, D.-Q., Jinadasa, K. B. S. N., Gersberg, R. M., Liu, Y., Tan, S. K., and Ng, W. J. (2015) Application of constructed wetlands for wastewater treatment in tropical and subtropical regions (2000–2013). *Journal of Environmental Sciences*, 30, 30–46.

Zhang, D. Q., Jinadasa, K. B. S. N., Gersberg, R. M., Liu, Y., Ng, W. J., and Tan, S. K. (2014) Application of constructed wetlands for wastewater treatment in developing countries – A review of recent developments (2000–2013). *Journal of Environmental Management*, 141, 116–131.

About the Author

Dr. Hamidi Abdul Aziz is a professor in environmental engineering in the School of Civil Engineering and the Head of the Solid Waste Management Cluster, Universiti Sains Malaysia. Professor Aziz received his PhD degree in civil engineering (environmental engineering) from the University of Strathclyde in Scotland in 1992. He is the Editor-in-chief of IJSRES, and the Managing Editor of IJEWM and IJEE. Professor Aziz's research focuses on alleviating problems associated with water pollution issues from industrial wastewater discharges and from solid waste management via landfilling, such as landfill leachate. Advanced oxidation processes is one of his research focuses.

In: Leachates
Editor: Maria R. Mattox
ISBN: 978-1-53612-508-5

Chapter 4

THE INDUCTION OF MICRONUCLEI AND NUCLEAR ABNORMALITIES IN PERIPHERAL ERYTHROCYTES OF *CLARIAS GARIEPINUS* (BURCHELL, 1822) FOLLOWING EXPOSURE TO SAWMILL WOOD LEACHATE

***Adekunle A. Bakare*[1,*], *Chibuisi G. Alimba*[1], *Titilayo Hassan*[2], *Ebenezer O. Ajayi*[2] and *Akindayo A. Sowunmi*[3]**

[1]Cell Biology and Genetics Unit, Department of Zoology, University of Ibadan, Ibadan, Oyo State. Nigeria
[2]Ecology and Environmental Biology Unit, Department of Zoology, University of Ibadan, Ibadan, Oyo State. Nigeria
[3]Hydrobiology and Fisheries Unit, Department of Zoology University of Ibadan, Ibadan, Oyo State. Nigeria

[*] Corresponding Author Email: adekunle.bakare@ui.edu.ng and adebakar19@yahoo.com.

ABSTRACT

With the rising global human population, trees are increasingly being cut to produce furniture, paper and building materials for human use. Common phenomena are the piling up of logs of wood and sawmills' shavings in the environment. The logs of wood and sawmills' shavings generate leachate that contain xenobiotics capable of contaminating aquatic and terrestrial ecosystems. They may also induce adverse effects on biota. Information is scarce on the DNA damaging potential of sawmill wood leachate (SWL) in aquatic vertebrates. Herein, the 96 h median lethal toxicity and cytogenotoxicity of a SWL from a sawmill dumpsite at Ibadan, Nigeria in *Clarias gariepinus* were investigated. Six concentrations (10 - 100%) of the leachate were examined for 96 h acute toxicity. Fish were exposed to sub-lethal concentrations (1.32 - 21.05%) of the 96 h LC_{50} for 7, 14 and 28 days to assess micronucleus (MN) and nuclear abnormalities in peripheral erythrocytes of five fish per concentration. The 96 h LC_{50} of 42.09% was 1.54 times more toxic than the 24 h LC_{50} (64.98%), indicating that the toxicity of the sawmill leachate increased with exposure time. The leachate induced significant ($p<0.05$) time-dependent increase in the frequency of MN and other nuclear abnormalities such as bleb, notch, bud and binucleated nuclei compared with the negative control. The concentrations of some of the SWL constituents were above permissible limit; and their interactions in the test organism provoked the observed mortality and genomic disruptions. These suggest that wood leachates can cause somatic mutation in fish.

Keywords: African catfish, aquatic conservation, DNA damage, hazardous substances, 96 h median lethal toxicity, micronucleus, wood wastewater

1. INTRODUCTION

Anthropogenic activities, which result in the discharge of solid wastes and wastewater into the environment, are fast increasing the deterioration of aquatic and terrestrial ecosystems. With the rising global human population, trees are increasingly being cut to produce finishing products such as furniture, paper and building materials for human use. Common phenomena are the piling up of logs of woods (many of which are treated with chemical preservatives) and sawmill shavings in the environment. These have elicited public health concern due to the production of dark brown leachate which may adversely affect aquatic ecosystem (Weis et al., 1995; Libralato et al., 2007; Ogunbode et al., 2013). Leachates generated from wood and sawmill activities are acutely toxic and deleterious to aquatic forms via deterioration of water qualities (Arimoro et al., 2007; Adeogun et al., 2011).

In Nigeria, wood shavings are directly discharged into lagoons and along river banks. This clogs fish gills leading to their death and reduces aquatic productivity via decrease light penetration (Nwankwo and Akinsoji, 1989). Wood leachate contains a significant spectrum of toxic inorganic and organic pollutants that can affect fish distributions and vulnerability to parasites (FAO, 1991). Acute toxicity is the most frequently used test to assess the ecotoxicity of sawmill shavings' leachate to organisms (Taylor et al., 1996; Libralato et al., 2007). This test is a preliminary tool that evaluates the overall toxicity of pollutants via mortality of the test organisms but it does not provide adequate information on the effects of pollutants on body physiology and genetic material of the organisms. Cytogenetic damage from the deleterious actions of environmental contaminants is of uttermost concern as it may lead to adverse reproductive damage, xenobiotic-induced stress syndromes including cancer formation and death of aquatic forms (Viarengo et al., 2007). It is hypothesized that the use of median lethal toxicity and cytogenotoxicity testing in the routine monitoring of sawmill wood leachate will provide comprehensive information on the general toxicity and possible genetic damage induced by the mixture of contaminants in the leachate.

Fishes are good model for monitoring acute toxicity and cytogenetic effects of aquatic pollutants. They are sensitive to the toxic effects of pollutants, metabolize and bioaccumulate chemicals in their bodies (Biuki et al., 2011) and serve as major route for human exposure to toxic substances in aquatic environment (WHO, 1990). Piscine MN test is a sensitive and suitable cytogenotoxic endpoint capable of detecting DNA and nuclear damage induced by xenobiotics via MN and abnormal nuclear formation in the erythrocytes of fishes (Al-Sabti and Metcalfe, 1995). The MN assay is one among cytogenetic biomarkers that strongly correlate with the deleterious effects of the mixture of contaminants measured in different bioindicators as well as in teleosts (Bolognesi and Hayashi, 2011; Harabawy and Mosley, 2014; Alimba and Bakare, 2016). *Clarias gariepinus*, the most locally consumed African clariid in Nigeria (Hayler, 1993), was used as the model organism in this study due to its rapid turnover rate and high adaptation to varying laboratory conditions. The suitability of this fresh water species in the cytogenotoxic monitoring of e-waste leachates and contaminated water (Bakare et al., 2013), landfill leachates (Alimba and Bakare, 2016) and abattoir effluent (Alimba et al., 2015) has been reported.

Despite information on potential deleterious effect of wood waste leachate over the years, there is paucity of information on genetic damage that may be induced by the complex mixture of toxic substances in the wood waste leachate. In this study, the 96 h median lethal toxicity and potential to induce micronucleus and nuclear lesions by sawmill wood leachate in *C. gariepinus* were investigated. Some physicochemical parameters and heavy metals were also analyzed in the leachate.

1.1. Wood Waste Management in Nigeria

Among different industrial activities that have been contributing waste to the environment, concerns have been recently raised on wood-based industries where forest products are used for various products such as building, mining, packaging, leather tannings amongst others in the country (Ogunbode et al., 2013). Round-wood in Nigeria comes mostly from the natural forest zone of the country, mainly from the Southern States: Cross River, Edo, Delta, Ogun, Ondo, Ekiti, Osun and Oyo States. The most important wood products consumed and traded in Nigeria are sawn-wood, plywood, particle board, news-print, printing and writing paper (Bello and Mijinyawa, 2010). The major wood processing industries in Nigeria are typically large capacity facility industries such as large sawmills, plywood mill, pulp and paper plants and quite many small-scale wood product manufacturing companies such as furniture industries, cabinet makers and carpentry. Majority of the saw mill industries are in the wood producing rain forest areas of the country in which the western states are among. The largest concentration of sawmills is in Lagos, Ekiti, Osun, Cross River, Ondo, Oyo, Imo, Edo, Delta and Ogun States. Together, they accounted for over 90% of the saw milling activities in the country. This indicated that guaranteed log supply is a major factor in the location of sawmills in Nigeria.

Sawmilling has been defined severally as a system, as an industry (Bennett, 1974) and as a process (Lucas, 1975). As a process, it involves converting log into lumber using different methods such as live-sawing (sawing around the log), slash cutting, and cant-sawing facility (Egbewole et al., 2011). In Nigeria, forest-based industries generate an impressive amount of waste during operational phases, from logging until the final product. According to Lasode et al. (2010) the quantity of wood waste generated is a function of the volume of planks offloading at various plank markets at any given time. About 50 – 60% of the log volume ends up as waste usually in the form of slabs, sawdust, edges and shavings. According to Dosunmu and Ajayi (2002) about 294,748 tons of wood waste is generated in Lagos state while about 2,288m^3 is generated daily in Abeokuta, Ogun State. Aina (2006) reported that a total of 104, 000m^3 wood wastes is generated daily in Nigeria. The situation is not very different in Ghana, a neighboring West African country, where wood residues amounted to 50% of which 8-10% is solid wood residues (Noack, 1995).

A breakdown of wood waste shows that about 20.64%, 10.28%, 7.3% and 6.31% comprise of slabs, sawdust, wane and bark respectively (Izekor and Kalu, 2008). Sawdust contributes the second largest proportion of wood waste generated from wood product operations. Sawdust, as the name implies, consists of fine particles of wood produced from cutting with a saw (Eze et al., 2011). It is the main product of sawmills, a by-product from agricultural, forestry or architectural activities and can serve many useful purposes such as mulch, fuel and sorbent material. In Nigeria, the huge amount of sawdust materials generated is commonly combusted, abandoned or thrown into rivers, streams or drainage. Despite the benefits derived from sawdust, if not adequately managed, it may constitute an environmental hazard. Common disposal methods include heaping at the mills, open air combustion, disposal along the roadside and into water bodies (Ogunbode et al., 2013).

1.2. Characteristics of Sawdust Leachate

In the presence of water, usually rainfall, run off or sprinkling water, wood leachate is produced when the liquid percolates through woodpiles, log yards, wood product storage areas, and wood waste disposal sites of various wood processing mills and barge loading facilities (Wendong et al., 2005). Sawdust derived from either coniferous (softwood) or deciduous (hardwood) or a combination of both, is made up of basic biodegradable organic chemicals such as cellulose, lignin, tannin, resin acids, terpenes, fatty acids (Hedmark and Scholz, 2008). The chemical composition of sawdust leachate varies with wood type, wood component (bark, heartwood, sapwood or foliage), waste compaction, leaching time and characteristics of leaching water (Alkassabeh et al., 2009).

1.2.1. pH

The pH of leachate influences the chemical process; it is an indicator of aggressiveness, aerobic and anaerobic condition of the leachate. Low pH level has been reported in wood leachate. Wendong et al. (2005) reported pH of 4 in wood leachate and they attributed it largely to some forms of coloured weak acids, such as the tannic acids that undergo a shift in their protonation. Adeogun et al. (2011) similarly observed that upper Ogun River at the discharge point of sawmills' effluents had acidic pH. They observed that the acidity maybe as a result of humic acid formed from decaying organic matter from sawdust leachate.

1.2.2. Cellulose and Lignin

Wood is made up of cellulose and lignin. Lignin makes up about a third of the mass of typical wood and adds great strength to the wood. It is a complex set of aromatic molecules and a polymer with a complex molecular structure. It is the major constituent of plant secondary cell wall and accounts for about 10 to 35% of total plant dry mass. It functions to keep various cellulose layers in lignocellulose materials together (Sun and Cheng, 2002). Cellulose is a polymer of sugars (Anderson et al., 2007). Cellulose fibrils have high tensile strength that is used in the textile industry, paper and miscellaneous materials like vulcanized fibre, plastic filters, filtering media and surgical cotton. Other uses include adhesives, explosives, thickening agents, coated paper, cellophane, artificial leather, films and foils (Hitchner and Leatherwood, 1982).

1.2.3. Tannin

Tannin is a naturally occurring water soluble polyphenolic compound commonly found in higher herbaceous and woody plants. Tannins function as a defence to stop fungi and bacteria from attacking the wood (Chung et al., 1998). Both tannins and lignin are hard to biodegrade (Tao et al., 2005). The tannins in wastewater generated from the removal of the bark from the log (debarking) in wood processing have been found to contribute to the toxicity of the waste stream (Field et al., 1991). Tao (2005) investigated the characteristics of leachate from a wood waste (sawdust) Pile in Columbia and revealed high values of tannin (3066–5150 mg L^{-1}) as tannic acid.

1.2.4. Chemical Oxygen Demand and Biochemical Oxygen Demand

COD measures the amount of the oxygen needed to oxidize all organic compounds in the sample, while the BOD measures the amount of oxygen uptake under a limited time by microbial activity. Svensson (2010) reported significant high leaching values of COD (> 4 000 mg/L) from a sawdust leachate. Steele (2004) reported that oak wood waste leachate North Ontario has a biochemical oxygen demand extremely greater than the required limit of 250 mg/L in water. Also, McDougall (1996) reported high levels of BOD from log yard run off in Alberta for a period of two years. In addition, Arimoro et al. (2007) recorded high value of BOD in Ethiope River receiving sawdust effluent.

1.2.5. Resin Acids

These are also toxic compounds found in wood. The different types of resin acids are part of the tree's defence system against insects and microorganisms (Keeling et al., 2006). Softwoods contain high concentrations of resin acids such as dehydroabietic acid (DHA), abietic acid, pimaric acid, and isopimaric acid, etc. (Wang et al., 1994). However, the concentrations vary within a tree and between tree species. Similarly, the types and concentrations of resin acids in effluents vary substantially, depending on the wood species, pulping and bleaching processes, and biological treatments (Taylor, 1988). Mcdougall (1996) analyzed samples of leachate from stagnant ponded water receiving wood waste (sawdust) from five saw mills in Alberta and he reported values of 0.02 to 10.2 mg/L respectively for resin and fatty acids which were toxic to aquatic life.

1.2.6. Heavy Metals

Occurrence of heavy metals in wood products may result from water-soluble inorganic pesticides that are commonly used as wood preservatives to make it resistant to attack by termites and fungi that cause decay. Such pesticides include Chromated Copper Arsenate (CCA). During treatment of wood with the pesticides, residues can seep in and leach CCA onto the wood surface in the presence of water directly retaining some of its constituents in the wood (CDPH, 2007). CCA's wood preserving chemicals are three metal oxides: copper oxide (CuO), arsenate (As_2O_5), and chromate (CrO_3). The purpose of copper is to serve as a fungicide, arsenate serves as an insecticide, and chromate is used to fix the other two metals to the wood. However, its use is being restricted in some countries, due to reported toxic effects on the environment. Mobility and biological availability of heavy metals depend not on their total concentration but on the physicochemical form in which they occur (Filgueiras et al., 2002, Smichowski et al., 2005). Mcdoguall (1996) reported high values of heavy metals (chromium, copper, arsenic) analyzed in surface water run off samples receiving sawdust leachate from two wood treatment facilities at Spray Lake Sawmills and Sunpine Forest Industries in Alberta and he attributed the occurrence of these metals to the use of wood preservative to prevent the wood log from termites' attack.

Furthermore, uncontrolled burning of sawmill wastes in open air usually produce smoke because of incomplete combustion in which most of the heavy metals in the wood bark and other wood waste are retained in the ash (Poykio et al., 2008).

1.3. Toxicity of Sawdust Leachate

Sawdust leachate has been reported to contain biodegradable organic matter, which increases the organic loads and turbidity, reduces dissolved oxygen content of the recipient water body and emits foul odour of hydrogen sulphide. It further enhances fungal and bacterial populations. All these render the receiving water body unfit for aquaculture, domestic and other uses. There are limited studies on the toxicity of sawdust leachate. In a study by Taylor et al. (1996), Aspen wood leachate toxicity to aquatic life was evaluated. Piles of aspen *(Populus tremuloides)* logs, harvested in winter and stock piled in wood yards until spring were observed to produce a dark, watery, acutely toxic leachate in waterways in Canada. Fresh aspen leachate derived from a 1:9 wood-water mixture (35 days immersion) was characterized by amber colour, low pH, extremely high BOD and high conductivity. The leachate was rich in phenols, organic carbon and organic nitrogen. Median acutely toxic concentrations of the leachate were consistently 1 to 2% of full strength for trout and *Daphnia.* Inhibition of bacterial metabolism began at concentrations below 0.3%. Leachate was less toxic to plant life but inhibited growth at concentrations of 12 to 16%. Toxicity of aspen leachate persisted at the same level as in fresh leachate for more than 2 months unless artificial aeration was provided. The authors believed oxygen depletion, low pH and phenolic compounds contribute to the toxicity of aspen leachate, but much of the toxic effect may be attributed to other unidentified constituents.

Another study was undertaken to demonstrate the effects of a 10-day exposure to sediments mixed with leachate from marine piles made from Southern pine *(Pinus southern*) that were treated with chromated copper arsenate Type-C[CCA-C] to a retention of 40 kg/m^3 (2.5pcf) and untreated Southern pine piles (Baldwin et al. 1996). The leachate was tested on the benthic amphipod, *Ampelisca abdita*. The biological endpoint used to establish effects was organism survival. Leachate obtained during a 28-day period was dosed onto both low organic carbon and high organic carbon sediments. The test organisms were exposed to each type of sediment at concentration of leachate ranging from 10 to 100%. The 10-day exposure was maintained under static conditions with continuous lightning to ensure maximum exposure to sediment. Results showed that leaching of copper, chromium and arsenic from properly treated and fixed CCA-C treated wood does not occur at concentrations that would adversely affect the survival of test organisms. Leachate from the treated piles had no significant effect on water quality.

According to Akiyama et al. (2001), tannic acids commonly found in herbaceous and woody plants show antimicrobial action and toxicity to filamentous fungi, yeast, bacteria and to certain viruses. In addition, they also have anti-nutritional and toxic effects such as reduced feed intake, growth rate, feed efficiency and net metabolizable energy in livestock (Chung, 1998). Fabio et al. (2012) revealed the toxicity of two different wood waste leachate-*Pinus sylvestris* (pine) and *Quercus robur* (oak) on freshwater microalgae (*Desmodesmus subspicatus*) and concluded that both wood species caused growth inhibitory effects and that oak was more toxic than pine. Food and Agricultural Organisation (FAO, 1991) showed that wood shavings and leachates are sources of inert solids and toxic pollutants that directly clog fish gill and indirectly reduce light penetration (Nwankwo and Akinsoji, 1989) thereby limiting productivity, contaminating the aquatic environment and making aquatic organism vulnerable. Furthermore, FAO (1991) noted that fishes exposed to wood leachate have weakened immune system leading to increased incidence of parasites. Wood wastes deposited into or near a water course can alter, disrupt or destroy fish habitat by smothering spawning ground areas and thus decreasing fish variety and abundance (FAO, 1991). Schaumburg (1973) studied leachate from short floating logs under laboratory conditions and measured various parameters including chemical oxygen demand (COD), biochemical oxygen demand (BOD) and acute toxicity. He found that the colour producing substances were primarily associated with leachate from the bark and the leachate was toxic to fish.

Arimoro et al. (2006) investigated the impact of sawmill wood wastes on the water quality and fish communities of Benin River, Niger Delta area of Nigeria. The study showed that prominent features of sawmill wood wastes had a negative effect on fish distribution and their communities. This effect may have resulted from the high chemical and particulate materials present in the wood wastes, which provided suitable location for high bacterial and fungal biodegradation activities. The sawmill wood wastes also impacted adversely on water quality. Olaitan (2010) investigated the histopatological and heamatological effects of sawdust leachate on mice. The result showed pathological changes in the organs (liver and kidney) and in the neurophils and lymphocytes of the exposed mice when compared with the control group. Similarly, Adeogun et al. (2011) reported the impact of abattoir and sawmill effluents on water quality of upper Ogun River in Nigeria. Elevated levels of physico-chemical parameters showed that the water quality of Upper Ogun River was adversely affected and impaired by effluents discharged from the abattoir and saw-mill industries especially during the dry season. Rex et al. (2016) assessed leachate toxicity and chemistry generated from pure aspen (*Populus tremuloides* Michx.), lodgepole pine (*Pinus contorta* Dougl.), hybrid white spruce (*Picea engelmannii* x glauca Parry) and black spruce (*Picea mariana* (Mill.) Britton). The leachates generated using rainfall simulation (a static 28-day laboratory assay) and a field-based exposure were analyzed for organic matter content, phenols, ammonia, pH, and toxicity using the luminescent bacterium *Vibrio fischeri* (Microtox™ analysis). Regardless of the treatment methods, the leachates contained high organic matter which increased toxicity in *V. fischeri*.

2. Materials and Methods

2.1. Study Site and Wood Leachate Collection

The study site, Bodija Sawmill (Longitude 7°25′ 35 ″N and Latitude 3°54′ 39″E), is in Ibadan, Oyo State, Nigeria. The sawmill activities (Figure 1A) generate about 140 tons of wood shavings daily and 32,935 tons annually (Ogunbode et al., 2013). The wood shavings are heaped up (Figure 1B) as solid wastes around residential quarters and are occasionally set on fire to reduce the volume. During the rainy season, raw leachate is usually generated from the heaps when water percolates through the shavings and this is transported as run off into nearby aquatic ecosystem. Leachate was collected from within the heaps of wood shavings (Figure 1C) at five different points in the sawmill into a clean transparent 25 L plastic container to give homogenous mixture from the site. It was then transported to the Ecology and Environmental Biology laboratory in the Department of Zoology, University of Ibadan, where it was sieved (0.7 mm sieve) to remove wood debris. The filtrate, considered as stock solution (100%), was designated sawmill wood leachate (SWL). It was kept at room temperature until use for the bioassays.

Figure 1. Sawmill operation at Bodija plank market in Ibadan, Nigeria (A); sawdust dumpsite (B) and leachate well (C) in the market.

2.2. Physicochemical Parameter and Heavy Metal Analyses of Wood Leachate

The sawmill wood leachate and tap water (negative control) were analysed for some physico-chemical parameters: chemical oxygen demand (COD), biochemical oxygen demand (BOD), dissolved oxygen (DO), conductivity, chloride, sulphate, nitrate, pH, temperature, lignin and tannin, and heavy metals: lead (Pb), cadmium (Cd), arsenic (As), chromium (Cr) and copper (Cu) in accordance with standard methods (USEPA, 1996; APHA, 1998). The concentrations of the metals in the samples were estimated using PerkinElmer A3100 atomic absorption spectrophotometer.

2.3. Animal and Experimental Design

Juvenile *C. gariepinus* (mean ± SD body weight; 8.62 ± 2.40 g and length; 9.20 ± 1.20 cm) obtained from the Fisheries Department of Oyo State Ministry of Agriculture and Natural Resources, Ibadan, were utilised. They were acclimated to laboratory conditions of 26 ± 1 °C and 12/12 h dark/light modes for 14 days prior to the experimental set-up. They were stocked at a population density of 15 fish per 25 L transparent plastic aquarium containing dechlorinated water and fed 5% of their body weight with standard fish feed (Durante 2 mm floating feed®), containing 35% crude protein twice daily.

2.4. Acute Toxicity Bioassay and Assessment of Quantal Response (Mortality)

This was in accordance with Organization for Economic Cooperation and Development (OECD, 1992). After range finding test, six concentrations: 10, 20, 40, 60, 80 and 100%, (v/v, SWL/dechlorinated tap water) of the leachate and control group (dechlorinated tap water) were set up in a static bioassay for the 96 h acute toxicity testing. Ten juvenile fish were stocked per group in duplicates and were not fed during the assay. Mortality assessment and behavioural changes were recorded every 24 h for the period of 96 h to determine the LC_{50} concentrations. Fish was assumed dead when there was no body or operculum movement, even when prodded with a blunt glass probe.

2.5. Exposure to Sub-Lethal Concentrations and Micronucleus Analysis

Fifteen juvenile *C. gariepinus* randomly selected into 25 L tanks were exposed to 21.05, 10.52, 5.26, 2.63, and 1.32% (v/v WSL/dechlorinated tap water) sub-lethal concentrations (corresponding to 1/2, 1/4, 1/8, 1/16 and 1/32 respectively, of the determined 96 h LC_{50}) of the leachate by immersion for 7, 14, and 28 days. Five fish were analyzed at each sampling time from each concentration of the leachate. Similar treatments were given to fish exposed to dechlorinated tap water and 0.02 mL/L of benzene (Sigma, St. Louis, MO, Australia) as negative and positive controls respectively. The test leachate and the controls were replaced every 48 h to reduce accumulation of metabolic wastes, remains of food particles, and volatilization of less stable substances in the leachate and benzene. Both the control and treated fish were exposed in the same seasonal period to avoid the influence of seasonal fluctuations of abiotic conditions on the expression of MN and nuclear abnormalities (Strunjak-Perovic et al., 2009).

At the end of each exposure time, blood was collected from the caudal vein of each fish in the test and control groups for MN analysis. Thin smear of the peripheral blood was made on three pre-cleaned slides per fish. The slides were air dried, fixed in absolute methanol for 30 minutes, and counterstained with 10% May-Grunwald and 5% Giemsa (Alimba and Bakare, 2016). 3000 erythrocytes per fish were scored for micronucleus induction at 1000x. Other nuclear abnormalities (NAs) such as binucleated cells, blebbed nucleus, notched nucleus, budding nucleus and cells with intact cell and nuclear membranes were also scored as cytotoxic parameters (Carrasco et al., 1990) at the same magnification.

2.6. Statistical Analysis

Data on mortality were analyzed as LC_5 (lethal concentration that caused 5% mortality), LC_{50} (lethal concentrations that caused 50% mortality) and LC_{95} (lethal concentration that caused 95% mortality) at their corresponding 95% confidence interval by probit analysis (Finney, 1971) using SPSS™ version 17.0. Toxicity factor (TF) for 24 hourly relative potency measurements of the leachate was determined. Data for the MN and NAs were presented as mean ± SE (standard error). One-way ANOVA was used to determine the differences among various groups according to exposure duration, while Dunnett multiple post hoc test was used to compare the level of significance ($p < 0.05$) of each treated group with the negative control. Statistical analyses for the MN and NAs were conducted using Graphpad prism 5.0® computer program.

3. Results

3.1. Physicochemical Parameter and Heavy Metal Analyses of Sawmill Wood Leachate

The physicochemical parameters and heavy metals analyzed in the wood leachates are presented in Table 1. The dark brown and pungent smelling wood leachate had pH that is within the national and international allowable limits. Alkalinity, acidity, electrical conductivity, nitrate, TS, BOD, COD, and As and Cu concentrations were higher than respective values in the negative control, NESREA and USEPA allowable limits for effluent quality criteria standards. Tannin and lignin quantified in the leachate, were not detected in the negative control. Pb, Cd, Ni, and Cr were below detectable limit, while chloride, DO, and phosphate were below permissible limits by NESREA (Nigeria) and USEPA (USA) standards.

Table 1. Physicochemical parameters and heavy metals analyzed in sawmill wood leachate (SWL) and tap water

Parameters	Tap water	SWL[a]	NESREA[b]	USEPA[c]
Colour	Colourless	Dark brown	-	-
pH	8.2	8.0	6.5-8.5	6.5-8.5
[d]EC	44	136	-	1000
[e]DO	7.7	1.6	-	-
[f]BOD	0.78	601.3	3	250
[g]COD	1.69	811.4	30	410
Total solid	98.4	889	-	500
Alkalinity	3.06	72.1	-	20
Acidity	1.51	35.06	-	-
Chloride	56.7	12.2	300	250
Phosphate	0.2	1.3	3.5	5
Nitrate	0.2	60.3	9.1	10
Tannin	BDL	0.60	-	-
Lignin	BDL	0.78	-	-
Lead	BDL	BDL	0.01	0.015
Copper	BDL	0.60	0.001	0.30
Chromium	BDL	BDL	0.01	0.1
Arsenic	BDL	0.88	-	0.05
Cadmium	BDL	BDL	0.005	0.005
Nickel	BDL	BDL	0.01	-

All values are in mg/L except pH, alkalinity, acidity, and EC (μS/cm). BDL, below detectable limit.

[a] SWL (Sawmill wood leachate).

[b] NESREA: National Environmental Standards and Regulations Enforcement Agency (2011) (Nigeria) maximum permissible limits for effluent for fisheries and recreational quality criteria standards.

[c] USEPA: United States Environmental Protection Agency (2006). (http://www.epa.gov/safewater/mcl.html).

[d] Electrical conductivity.

[e] DO: Dissolved oxygen.

[f] BOD: biochemical oxygen demand.

[g] COD: Chemical oxygen demand.

3.2. Acute toxicity and Assessment of Quantal Response (Mortality)

Varying behavioral responses ranging from erratic swimming movements using lateral and ventral sides to darting up and down with occasional jumpy movements were observed in exposed fish. Frequent opercula movements were also common among treated fish. Fish in the exposed groups became weak and settled at the bottom of the tank and died. Mortality in the exposed fish was concentration-dependent with higher frequency of mortality recorded at the 40 (15%), 60 (70%), 80 (90%) and 100 (100%)% concentrations of SWL. There were no adverse reactions and mortality in the control set up during the period of exposure. The derived toxicity indices from the concentration-mortality data of SWL ranged from 64.98% (24 h LC_{50}) to 42.09% (96 h LC_{50}). The toxicity factor computed from the 96 h LC_{50} showed that SWL was highly toxic to *C. gariepinus* during the 96 h LC_{50} (TF = 1.54). The derived LC_5, LC_{50} and LC_{95} data for the 24 – 96 h acute toxicity showed that *C. gariepinus* mortality increased with SWL concentration and exposure duration (Table 2).

Table 2. 96 h acute toxicity determination of sawmill wood leachate (SWL) using *Clarias gariepinus*

Time (h)	LC_5 (95% CI)	LC_{50} (95% CI)	LC_{95} (95% CI)	Slope ± S.E	Probit line equation	TF[a]
24	35.93(-1.25–1.74)	64.98 (0.92–1.91)	117.54 (1.97–3.20)	6.39 ± 2.88	Y = -6.58 + 6.39X	1.00
48	31.29 (0.67–1.67)	54.89 (1.41–1.84)	96.31 (1.88–2.28)	6.74 ± 2.35	Y = -6.72 + 6.74X	1.18
72	30.45 (0.81–1.63)	47.99 (1.40–1.78)	75.64 (1.79–2.13)	8.32 ± 2.90	Y = -8.99 + 8.32X	1.35
96	25.29 0.99–1.54)	42.09 (1.45–1.72)	70.05 (1.75–2.07)	7.43 ± 2.13	Y = -7.08 + 7.43X	1.54

[a] TF = Toxicity factor derived from 96 h LC_{50}; CI= 95% confidence limit; S.E = Standard error.
LC_5: Lethal concentration that caused 5% mortality of exposed *C. gariepinus* at 95% CI.
LC_{50}: Lethal concentration that caused 50% mortality of exposed *C. gariepinus* at 95% CI.
LC_{95}: Lethal concentration that caused 95% mortality of exposed *C. gariepinus* at 95% CI.

3.3. Micronucleus and Abnormal Nuclear Formation in Peripheral Erythrocytes of *C. gariepinus*

The tested wood leachate induced concentration-dependent significant ($p < 0.05$) increased in the frequency of micronucleated erythrocytes (Figure 2) in peripheral blood of *C. gariepinus* at the different exposure durations. The induced MN in the wood leachate treated *C. gariepinus* was higher than that in the negative control group by varying fold increase and this depended on concentration and exposure duration (Figure 3). Table 3 presents data on the frequency ($p<0.05$) of nuclear abnormalities (Figure 2) induced in peripheral erythrocytes of *C. gariepinus*. The induction of other nuclear abnormalities was not concentration-dependent. It decreased with increasing SWL concentration, but was higher at the 7 and 28-day exposure periods.

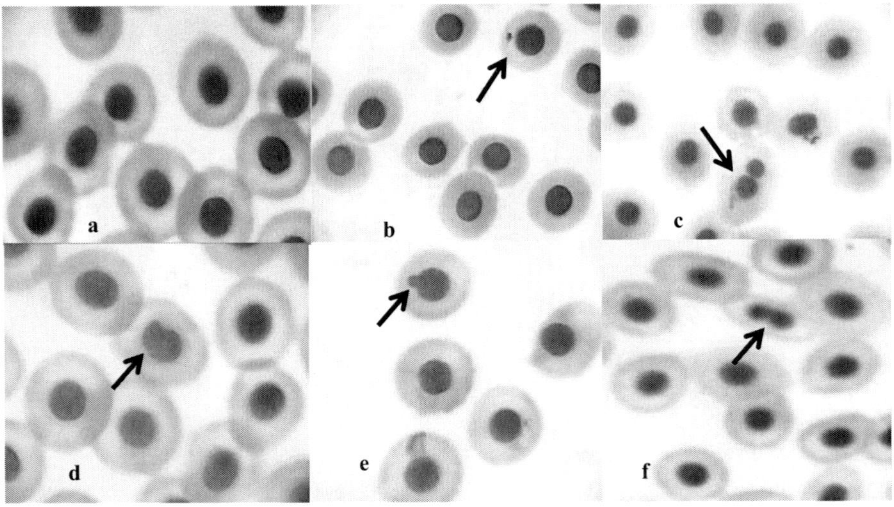

Figure 2. Micronucleated and nuclear abnormalities observed in the peripheral blood of wood leachate treated *Clarias gariepinus*. (a) normal peripheral erythrocyte. (b) micronucleated peripheral erythrocyte. (c) binucleated peripheral erythrocyte. (d) peripheral erythrocyte with blebbed nucleus. (e) peripheral erythrocyte with budded nucleus. (f) peripheral erythrocyte with notched nucleus (x1000).

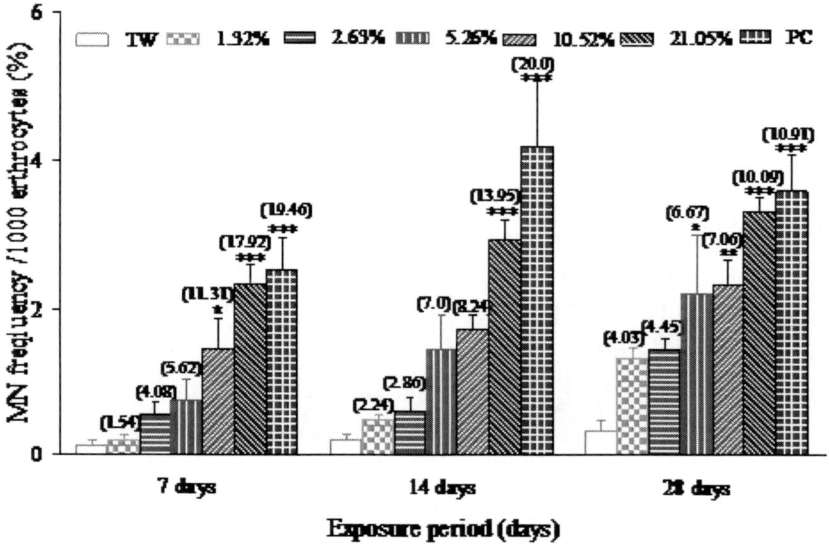

Figure 3. Frequency of micronucleus formation in peripheral erythrocytes of *Clarias gariepinus* exposed to wood leachate. $^{*}p < 0.05$; $^{**}p < 0.01$; $^{***}p < 0.001$ which are significantly different from corresponding negative controls (tap water) using Dunnett's multiple post hoc comparison test. Values in parentheses are fold increase from the corresponding negative control. PC= positive control (Benzene (0.02 mL/L).

Table 3. Total nuclear abnormalities (NAs) in peripheral erythrocytes of *C. gariepinus* exposed to wood leachate (SWL)

Treatment	Conc (v/v;%)	7 days	14 days	28 days
Tap water	0	0.05 ± 0.03	0.05 ± 0.03	0.07 ± 0.03
Benzene	0.02 (mL/L)	0.10 ± 0.03	0.05 ± 0.03	0.12 ± 0.03
SWL	1.32	0.38 ± 0.11^{a}	0.31 ± 0.10^{a}	0.51 ± 0.06^{a}
	2.63	0.31 ± 0.06^{a}	0.39 ± 0.05^{a}	0.22 ± 0.02^{a}
	5.26	0.35 ± 0.04^{a}	1.26 ± 0.04^{a}	0.11 ± 0.01
	10.52	0.08 ± 0.04	0.26 ± 0.03^{a}	0.27 ± 0.09^{a}
	21.05	0.10 ± 0.03	0.07 ± 0.04	0.43 ± 0.15^{a}

End points represent Mean ± SE/1000 cells. Superscripts are significantly ($^{a}p < 0.05$) different from negative control using Dunnett's multiple post hoc comparison test.

Discussion

Rapid chemical pollution of the aquatic environment is altering natural conditions of the water bodies leading to functional imbalance in aquatic organisms. There is increasing concern that the spectrum of inorganic and organic constituents in sawmill wood leachates may adversely increase cytogenetic damage in aquatic forms. High concentrations of electrical conductivity, TS, phosphates, nitrates, BOD, and COD and low concentration of DO recorded in the SWL compared with the negative control and permissible standards are in concert with those reported from upper Ogun River (Adeogun et al., 2011) and Benin River (Arimoro et al., 2007) in Nigeria, which received untreated sawmill wood leachates. These physico-chemical parameters are capable of inducing toxicity on aquatic biota and altering water parameters of aquatic habitats. The dark brown colour of the leachate can obscure visibility and reduce the aesthetic values of aquatic environments. Low turbidity (transparency) and reduced oxygen solubility in water bodies were linked to sawmill wood leachate (Jonnalagadda and Mhere, 2001; Arimoro et al., 2007). Low DO in the tested wood leachate suggests increased anaerobic condition which can impair respiration in aquatic species (Victor and Tetteh, 1988). Low DO showed oxygen deficit and corroborated high BOD and COD concentrations suggesting high organic compounds and hydrogen sulphide formation in the tested leachate (Arimoro et al., 2007). These conditions might have contributed to the deleterious effects of the tested leachate on *C. gariepinus*.

The 96 h LC_{50} of 42.09% showed that the tested leachate was toxic to *C. gariepinus*. High total solids of the leachate possibly contributed to the dose-responsive increased mortality of *C. gariepinus*. The inert solids from the wood shavings probably directly clog fish gills leading to the death of the fish (FAO, 1991). They can also indirectly reduce light penetration in water bodies that will reduce aquatic productivity (Nwankwo and Akinsoji, 1989; FAO, 1991). It is also possible that the analyzed physicochemical parameters and toxic metals interacted synergistically, antagonistically or additively to induce the elicited toxic effects in *C. gariepinus* (Cameron and Koch, 1980; Cheung et al., 1993). The 96 h LC_{50} of 42.09% was 1.54 times more toxic than the 24 h LC_{50} (64.98%) indicating that the toxicity of the SWL increased with exposure time. Studies have shown that trembling aspen (*Populus tremuloides* Michx.) wood leachate induced median acute toxicity of 1–2% to trout and *Daphnia,* inhibited bacterial metabolism at >0.3% and algal growth at 12–16% concentrations (Taylor et al., 1996). Furthermore, wood leachates from different wood without preservatives induced median acute toxicity of 9.0 (8.29–9.78) μg/L to oyster (*Crassostrea gigas*) and 5.1 (4.6–5.6) mg/L to brine shrimp (*Artemia franciscana*) (Libralato et al., 2007). Different aquatic organisms, even in their life cycles, are not equally susceptible to the same toxicants (ECETOC 2005). Information is relatively scarce on the median acute toxicity of sawmill wood leachate to African catfish (*C. gariepinus*). In comparison with previously reported 96 h LC_{50}'s of a landfill leachate (Alimba and Bakare, 2016) and abattoir effluent (Alimba et al., 2015) in *C. gariepinus*, the value obtained herein suggests that African catfish seems to be more tolerant to the tested SWL.

The sub-lethal concentrations of the sawmill leachate when released into the aquatic environment may not result in immediate mortality of aquatic biota. However, it is possible that bioaccumulation of the constituents especially the toxic metals (Cu and As) and other constituents over a period of time may cause increasing health hazards in aquatic organisms and higher trophic level organisms including humans. The cytotoxicity and genotoxicity induced in *C. gariepinus* by the sub-lethal concentrations of the SWL via increased MN and abnormal nuclear formation in peripheral erythrocytes showed that the SWL constituents are clastogenic and/or aneugenic. Increased MN and NA indicate genetic alterations which are related to somatic mutation and cancer formation in fish species (Torres de Lemos et al., 2007).

Instability in fish genome may lead to biodiversity loss via decrease embryonic viability, genetic related disorders and reduced fitness (Kurelec, 1993). Arsenate analyzed at higher concentration in the tested leachate induces clastogenic and aneugenic effects via oxidative DNA adduct and DNA-protein cross-links leading to mutation and carcinogenesis

(Bhattacharjee et al., 2013). Similarly, the observed high concentration of copper in the leachate may induce DNA damage by binding to sensitive molecules and eliciting dangerous free radicals (Gabbianelli et al., 2003). It is also possible that high nitrogenous compounds in the leachate (as indicated by high nitrates) formed reactive free nitric acid (NO), a cytotoxin, with mutagenic and carcinogenic properties, which increased the induction of micronucleated erythrocytes in the exposed *C. gariepinus* (Burney et al., 1999). The induction of cytotoxicity and genotoxicity in *C. gariepinus* herein may involve free radical formation. This assertion is in accord with previous observation/report wherein e-waste leachate induced MN and NAs in *C. gariepinus* via induction of oxidative stress (Bakare et al., 2013). Lignins and tannins analyzed in the sawmill leachate are water soluble polyphenols usually found in many plants. They contribute mostly to the toxicity of wood leachates in aquatic species (Bailey et al., 1999; Taylor and Carmichael, 2003), but information relating to their cytogenotoxic and mutagenic effects remained unclear (Chung et al., 1998). Dose and time dependent variations in the MN and NAs formations are in accordance with similar trend in eastern mud minnow, Umbra pygmaea, exposed to ethyl methanesulfonate (Hooftman and deRaat, 1982), and in *C. gariepinus* exposed to abattoir effluent (Alimba et al., 2015) and landfill leachates (Alimba and Bakare, 2016).

We have observed that other NAs were consistently formed along with MN in response to solid waste leachate exposure (Bakare et al., 2013; Alimba and Bakare, 2016), though at different frequencies and concentration-effect relationships as observed herein. Therefore, scoring NAs and MN as biomarkers of cytogenetic damage in *C. gariepinus* exposed to leachate is sensitive and efficient and augments the cytogenotoxic assessment of mixture of contaminants in leachates. Significant increase in total nuclear abnormalities herein showed that NAs are formed in response to sawmill leachate exposure. The time - independent expressions in the frequencies of MN and NAs observed in the SWL treated fish suggest that the peak time for the formation of MN and NAs erythrocytes in *Clarias gariepinus* occurs within 1–7 days after exposure to genotoxicants. This has been similarly observed when *C.*

gariepinus was exposed to mitomycin C (Bahari et al., 1994) and abattoir wastewater (Alimba et al., 2015). Furthermore, the higher frequencies of these biomarkers during the 28-day exposure also suggest that the spleen does not compromise the frequencies of MN and NAs in *C. gariepinus* since they have nonsinusal spleen (Udroiu, 2006). The constituents of the tested leachate possibly interfered with cell cycle and DNA synthesis in the fish via various mechanisms to elicit the observed increase in the frequencies of NAs (Serrano-Garcia and Montero-Montoya, 2001). For instance, the presence of binucleated cells (biomarker of cytotoxicity) suggests that the leachate constituents blocked cytokinesis of a normal dividing cell at M phase of the cell cycle (Fenech, 2007). The presence of notch, bud and bleb nuclei is associated with chromosome aneuploidy and terminal acentric fragmentation due to disturbance of the chromatin materials by xenobiotics in the leachate (Monteiro et al., 2011). The presence of bud and bleb nuclei formation has also been linked to the entrapment of extra-chromosomally amplified DNA during S-phase and is related to genotoxicity events (Fenech et al., 2011). The observed DNA damage induced by SWL may lead to genetic imbalance and carcinogenesis in aquatic forms (Mix, 1986; Malins et al., 1988; Rodilla, 1993).

CONCLUSION

Clarias gariepinus, the most locally consumed African clariid in Nigeria, was used as model organism herein to assess the cytogenetoxic effect of different concentrations of sawmill leachate from Bodija sawmill in Ibadan, Nigeria. The tested leachate contained particulate matters and chemical constituents which are capable of contaminating aquatic ecosystems, and adversely impacting on aquatic life. It induced a time-dependent acute toxicity and concentration-dependent significant increase in the frequencies of micronucleated peripheral erythrocytes and nuclear abnormalities in *C. gariepinus*. Xenobiotics in sawmill leachate are emerging clastogens and/or aneugens that are capable of inducing genomic instability in fish and other aquatic biota which can increase fish predisposition to health-related syndromes including tumour formation. Therefore, indiscriminate disposal of sawmill wood shaven into the terrestrial and aquatic environment may increase contaminants in the environment and adversely affect aquatic and terrestrial biota. It is therefore necessary that sustainable disposal methods be implemented by appropriate regulatory authorities to protect public and environmental health.

REFERENCES

Adeogun, A.O., Chukwuka, A.V., Ibor, O.R. 2011. Impact of Abbatoir and Sawmill effluents on water quality of upper Ogun River (Abeokuta). *American Journal of Environmental Sciences* 7 (6), 525–530.

Aina, O.M. 2006. Wood waste utilization for energy generation, proceedings of the International Conference on Renewable Energy for Developing Countries. http://cere.udu.edu/Aina.pdf accessed 20th August, 2014.

Akiyama, H., Kazuyasu, F., Yamasaki, O., Oono, T., Iwatsuki, K. 2001. Antibacterial action of several tannins against *Staphylococcus aureus*. *Journal of Antimicrobial Chemotherapy* 48 (48), 487-491.

Alimba, C.G., Ajayi, E.O., Hassan, T., Sowunmi, A.A., Bakare, A.A. 2015. Cytogenotoxicity of Abattoir effluent in *Clarias gariepinus* (Burchell, 1822) using micronucleus test. *Chinese Journal of Biology* volume 2015, Article ID 624524; 1-6. Doi.org/10.1155/2015/624524.

Alimba, C.G., Bakare, A.A. 2016. *In vivo* micronucleus test in the assessment of cytogenotoxicity of landfill leachates in three animal models from various ecological habitats. *Ecotoxicology* 25 (2), 310–319.

Alkassasbeh, J.Y.M., Heng, L.Y., Surif, S. 2009. Toxicity testing and the effect of landfill leachate in Malaysia on behavior of common Carp (*Cyprinus carpio* L., 1758; Pisces, Cyprinidae). *American Journal of Environmental Sciences* 5 (3), 209-217.

American Public Health Association (APHA). 1988. In *Standard Methods for the Examination of Water and Wastewater*, 20th ed.; APHA: Washington, DC, USA, 1988.

Al-Sabti, K., Metcalfe, C.D. 1995. Fish micronuclei for assessing genotoxicity in water. *Mutation Research* 343, 121–135

Arimoro, F.O., Ikomi, R.B., Osalor, E.C. 2007. The Impact of Sawmill wood wastes on the water quality and fish communities of Benin River, Niger Delta Area, Nigeria. *International Journal of Science and Technology* 2 (1), 1-12.

Anderson, N.A., Wanger, S.S., Schwandt, J.W. 2007. The *Plarsosus Ostreatu Sapious* species complex. *Journal of Mycologia* 6, 28–35.

Bahari, I.B., Noor, F.M., Daud, N.M. 1994. Micronucleated erythrocytes as an assay to assess actions by physical and chemical genotoxic agents in *Clarias gariepinus*. *Mutation Research* 313, 1–5.

Bailey, H.C., Elphick, J.R., Potter, A., Chao, E., Konasewich, D., Zak, J.B. 1999. Causes of toxicity in storm water runoff from sawmills. *Environmental Toxicology and Chemistry* 18, 1485–1491.

Bakare, A.A., Alabi, O.A., Gbadebo, A.M., Ogunsuyi, O.I., Alimba, C.G. 2013. *In vivo* cytogenotoxicity and oxidative stress induced by electronic waste leachate and contaminated well water. *Challenges* 4, 169–187.

Baldwin, W.J., Pasek, E.A., Osborne, P.D. 1996. Sediment toxicity study of CCA-C-treated marine piles. *Forest Products Journal* 46 (3), 42-50.

Bello, S.R., and Mijinyawa, Y. 2010. Bioenergy for Sustainable Development in Africa. Google Books. Accessed 20th August, 2014.

Bennett, W. 1974. Modern sawmill techniques. Proceeding of third sawmill clinic, Portland. Oregon. Miller Freeman Publications, Incorporation, California.

Bhattacharjee, P., Banerjee, M., Giri, A.K. 2013. Role of genomic instability in arsenic-induced carcinogenicity: a review. *Environment International* 53, 9–40.

Biuki, N.A., Savari, A., Mortazavi, M., Zolgharnein, H., Salamat, N. 2011. Accumulation and elimination of cadmium and lead in Juvenile milkfish during sublethal exposure. *Toxicology and Environmental Chemistry* 93, 2022–2033.

Bolognesi, C., Hayashi, M. 2011. Micronucleus assay in aquatic animals. *Mutagenesis* 26, 205–213.

Burney, S., Caulfield, J.L., Niles, J.C., Wishnok, J.S., Tannenbaum, S.R. 1999. The chemistry of DNA damage from nitric oxide and peroxynitrite. *Mutation Research* 424(1-2), 37–49.

Cameron, R.D., Koch, F.A. 1980. Toxicity of landfill leachate. *Journal of Water Pollution and Control Federation* 52, 761-769.

Carrasco, K.R., Tilbury, K.L., Myers, M.S. 1990. Assessment of the piscine micronucleus test as an in situ biological indicator of chemical contaminant effects. *Canadian Journal of Fisheries and Aquatic Sciences* 47 (11), 2123–2136.

Cheung, K.C., Chu, L.M., Wong, M.H. 1993. Toxic effects of landfill leachate on microalgae. *Water Air and Soil Pollution* 69(3-4), 337-349.

Chung, K.T., Wong, T.Y., Wei, C.L. 1998. Tannins in human health; a review. *Critical Review on Food* and *Science Nutrition* 38, 421-464.

Dosunmu, O.O., and Ajayi, A.B. 2002. Problems and management of sawmill waste in Lagos. Proceedings of International Symposium on Environmental Pollution Control and Waste Management pp 271-278.

Egbewole, Z.T., Ogunsanwo, O.Y., Omole. 2011. Technical efficiency of Lumber recovery from high Forest tree species in selected sawmills of Southwestern Nigeria. *Nigerian Journal of Agriculture, Food and Environment* 7 (1), 34-41.

European Centre For Ecotoxicology and Toxicology Of Chemicals (ECETOC) 2005. *Alternative Testing Approaches in Environmental Safety Assessment*. European Centre for Ecotoxicology and Toxicology of Chemicals, Brussels, Belgium. ECETOC Technical Report No. 97.

Eze, V.C., Uzoaru, N., Agwung-Fobellah, D. 2011. Isolation and characterization of microorganisms involved in degradation of sawdust waste in Rivers state, Nigeria. *Asian Journal of Science and Technology* 1 (4), 044-048.

Fabio, K., Paulo, S.S., Marcia, M., William, H. 2012. Toxicity of wood leachates from *Pinus sylvestris* and *Quercus robur* on the Microalgae *Desmodesmus subspicatus. Applied Science Research* 4 (2), 274-282.

Fenech, M. 2007. Cytokinesis-block micronucleus cytome assay. *Protocol* 2(5), 1084-1104.

Fenech, M.M., Kirsch-Volders, M., Natarajan, A.T., Surralles, J., Crott, J.W., Parry, J., Norppa, H., Eastmond, D.A., Tucker, J.D., Thomas, P. 2011. Molecular mechanisms of micronucleus, nucleoplasmic bridge and nuclear bud formation in mammalian and human cells. *Mutagenesis* 26, 125–132.

Field, J.A., Lettinga, G., Gary, W.H. 1991. Treatment and detoxification of aqueous Spruce Bark Extracts by *Aspergillus. Nigeria Water Science and Technology* 24, 127-137.

Filgueiras, A.V., Lavilla, I., Bendicho, C. 2002. Chemical sequential extraction for metal partitioning in environmental solid samples. *Journal of Environment Monitoring* 4, 823-857.

Food and Agricultural Organization (FAO). 1991. *African fisheries and the Environment.* FAO Regional Office, Accra, RAFR/91/02, Accra, Ghana 26.

Gabbianelli, R., Lupidi, G., Villarini, M., Falcioni, G. 2003. DNA damage induced by copper on erythrocytes of gilthead sea bream *Sparus aurata* and mollusk *Scapharca inaequivalvis. Archives of Environmental Contamination and Toxicology* 45(3), 350–356.

Harabawy, A.S.A., Mosleh, Y.Y.I. 2014. The role of vitamins A, C, E and selenium as antioxidants against genotoxicity and cytotoxicity of cadmium, copper, lead and zinc on erythrocytes of Nile tilapia, *Oreochromis niloticus. Ecotoxicology and Environmental Safety* 104, 28–35.

Hayler, G.S. 1993. Aspects of the biology and culture of the African catfish, *Clarias gariepinus*, (Burchell 1822) with particular reference to developing African countries. *Research Advances in Aquaculture IV*. Edited by Minor, J.F. and Roberts, R.J. pp 235–291.

Hedmark, Å., Scholz, M. 2008. Review of environmental effects and treatment of runoff from storage and handling of wood. *Bioresource and Technology* 99, 5997-6009.

Hitchner, E.V., Leatherwood, J. 1982. Use of cellulose depressed mutant of Cellulomonas in The production of single-cell-protein from cellulose. *Applied Environmental Microbiol*ogy 41, 465- 583.

Hooftman, N.R., de Raat, W.K. 1982. Induction of nuclear abnormalities (micronuclei) in the peripheral blood erythrocytes of the eastern mudminnow Umbra pygmaea by ethyl methanesulphonate. *Mutation Research* 104(1–3), 147–152.

Izekor, D.N., Kalu, C. 2008. Volume estimation of sawmill waste in Benin City Edo state, Nigeria. *Forest and Forest Product Journal* 1, 24–28.

Jonnalagadda, S.B., Mhere, G. 2001. Water quality of the Odzi River in the Eastern highlands of Zimbabwe. *Water Research* 35, 2371-2376.

Keeling, I., Bohlmann, J., Svensson, H. 2006. Diterpene resin acids in conifers. *Phytochemistry* 67, 2415-2423.

Kurelec, B. 1993. The genotoxic disease syndrome. *Marine Environmental Research* 35(4), 341–348.

Lasode, O.A., Balogun, A.O. 2010. Wood Waste Generation in Ilorin Metropolis: Problems Management and Prospects. Report of Twenty fifth International Conference on Solid Waste Technology and Management. *ICSW* 25 (3), 14 – 17.

Libralato, G., Losso, C., Volpi Ghirardini, A. 2007. Toxicity of untreated wood leachates towards two saltwater organisms (*Crassostrea gigas* and *Artemia franciscana*). *Journal of Hazardous Materials* 144, 590–593.

Lucas, E.B. 1975. Wasted Trees Products in Nigeria. Part I. Their causes extent and characteristics. *Nigerian Journal of Forestry* 5 (1 and 2), 24 – 30.

Malins, D.C., McCain, B.B., Landahl, J.T., Myers, M.S., Krahn, M.M., Brown, D.W., Chan, S.L., Roubal, W.T. 1988. Neoplastic and other diseases in fish in relation to toxic chemicals: an overview. *Aquatic Toxicology* 11, 43–67.

Mix, M.C. 1986. Cancerous disease in aquatic animals and their association with environmental pollutants: a critical literature review. *Marine Environmental Research* 20, 1–141.

Monteiroa, V., Cavalcante, D.G.S.M., Viléla, M.B.F.A., Sofia, S.H., Martinez, C.B.R. 2011. *In vivo* and *in vitro* exposures for the evaluation of the genotoxic effects of lead on the Neotropical freshwater fish *Prochilodus lineatus*. *Aquatic Toxicology* 104, 291–298.

NESREA. 2011. National Environmental (Surface and Groundwater quality) Regulations. National Environmental Standards and Regulations Enforcement Agency.

Noack, D. 1995. Making better use of tropical timber resources. *ITTO, Tropical Forest Update* 5 (2), 12-13.

Nwankwo, D.I., Akinsoji, S.O. 1989. The influence of sawmill wood wastes on diatom population at Okobaba, Lagos Nigeria. *Nigerian Journal of Botany* 11, 15-24.

Ogunbode, E.B., Fabunmi, F.O., Ibrahim, S.M., Jimoh, I.O., Idowu, O.O. 2013. Management of Sawmill Wastes in Nigeria: Case Study of Minna, Niger State. *Greener Journal of Science* 3 (2), 034-041.

Olaitan, B.A. 2010. Hitopathological and haematological effects of sawdust leachate on mice. M.Sc. project report, Department of Zoology, University of Ibadan, Ibadan, Nigeria, 65pp.

Organization for Economic Cooperation and Development (OECD) (1992). Guidelines for testing of chemicals. Guideline no. 203: fish, acute toxicity test.

Poykio, R., Ronkkomaki, H., Nurmesniemi, H., Tenno, T., Paama L. 2008. Extractability of heavy metals in cyclone ash. *Proceedings of the Estonian Academy of Sciences* 57, 100-106.

Rex, J., Dubé, S., Krauskopf, P., Berch, S. 2016. Investigating potential toxicity of leachate from wood chip piles generated by roadside biomass operations. *Forests* 7, 40.

Rodilla, V. 1993. Origin and evolution of binucleated cells and binucleated cells with micronuclei in cisplatin-treated CHO cultures. *Mutation Research* 300(3-4), 281–291.

Schaumburg, F.D. 1972. The influence of log handling on water quality. Oregon State University, Dept of Civil Eng. Project No. WPO 1320 for the Water Quality Office, Environmental Protection Agency, pp 115.

Serrano-Garcia, L., Montero-Montoya, R. 2001. Micronuclei and chromatid buds are the results of related genotoxic events. *Environmental and Molecular Mutagenesis* 38, 38-45.

Smichowski, P., Polla, G., Gomez, D. 2005. Metal fractionation of atmosphere aerosols via sequential chemical extraction. *Analytical Bioanalytical Chemistry ABC* 381, 302-316.

Steele, J. 2004. Wood waste processing and the environment. *Journal of Forestry* 5 (1), 2

Strunjak-Perovic, I., Coz-Rakovac, R., Popovic, N.T., Jadan, M. 2009. Seasonality of nuclear abnormalities in gilthead sea bream, *Sparus aurata* (L.) erythrocytes. *Fish Physiology and Biochemistry* 35(2), 287-291.

Sun, Y. and *Cheng*, J. *2002*. Hydrolysis of Lignocellulosic Materials for Ethanol Production: A Review. *Bioresource Technology* 83, 1-11.

Svensson, H. 2010. Leaching Test with Sawdust from Different Tree Species – Appropriateness of using them as adsorption media in wastewater and in storm water treatment. Master of science, Linnaeus University, 45p.

Tao, W., Hall, K., Masbough, A., Frankowski, K., Duff, S. 2005. Characterization of leachate from a woodwaste file. *Water Quality Research Journal of Canada* 40, 476-483.

Taylor, B.P., Yeager, K.L., Abernethy, S.G., Westlake, G.F. 1988. Scientific criteria document for development of provincial water quality objectives and guidelines: resin acids. Ontario Ministry of the Environment, *Water Resources Branch* 25 (3), 281-284.

Taylor, B.R., Goudey, J.S., Carmichael, N.B. 1996. Toxicity of aspen wood leachate to aquatic life: Laboratory studies. *Environmental Toxicology and Chemistry* 15 (2), 150–159.

Taylor, B.R., Carmichael, N.B. 2003. Toxicity and chemistry of aspen wood leachate to aquatic life: field study. *Environmental Toxicology and Chemistry* 22, 2048–2056.

Torres, C., Rodel, P.M., Terra, N.R., D'Avila, N.C., Bernardo, E.B. 2007. River water genotoxicity evaluation using micronucleus assay in fish erythrocytes. *Ecotoxicology and Environmental Safety* 66, 391–401.

Udroiu I. 2006. The Mammalian Spleen. Homo Legens, Rome.

USEPA. 1996. Acid digestion of sediments sludges and soils. Method-3050B.USEPA: Washington, DC.

Viarengo, A., Lowe, D., Bolognesi, C., Fabbri, E., Koehler, A. 2007. The use of biomarkers in biomonitoring: a 2-tier approach assessing the level of pollutant-induced stress syndrome in sentinel organisms. *Comparative Biochemistry and Physiology* 146 (3), 281–300.

Victor, R., Tetteh, J.O. 1988. Fish communities of a perturbed stream in Southern Nigeria. *Journal of Tropical Ecology* 4, 49-59.

Wang, Z., Tao, C., Yong, G., Colette, B., Yasu, H. 1994. Biological Degradation of Resin Acids in Wood Chips by Wood-Inhabiting Fungi. *Journal of Enviromental Monitoring* 4, 823-857.

Weis, P., Weis, J.S., Couch, J., Daniels, C., Chen, T. 1995. Pathological and genotoxicological observations in oysters (*Crassostrea virginica*) living on chromated copper arsenate (CCA)-treated wood. *Marine Environmental Research* 39, 275-278.

WHO. 1990. Environmental health criteria for methylmercury, Environmental Health Criteria 101, Geneva. pp. 144.

BIOGRAPHICAL SKETCH

Adekunle A. Bakare

Affiliation: University of Ibadan, Ibadan, Nigeria

Education: BSc, MSc, PhD.

Business Address: Cell Biology & Genetics Unit, Department of Zoology, Faculty of Science, University of Ibadan, Ibadan, Nigeria.

Research and Professional Experience:

My major research focus is on genotoxicity and mutagenicity of chemical substances in the Nigerian environment. This is with the understanding that human exposure to these chemicals is almost inevitable; as these chemicals can induce DNA damage that may affect not only the present but also the future generations. I have studied the genotoxicity and mutagenicity of solid waste leachates, industrial effluents, pesticides, analgesics, medicinal plant extracts and electronic waste elutriate in Nigeria. I have extensively used in my investigations the *Allium cepa*, murine sperm morphology, mouse bone marrow chromosome aberrations and micronucleus, and the single cell gel electrophoresis (Comet assay) assays to show the potential capacity of these substances to induce genetic damage *in vitro* and *in vivo*. I have extended the use of some of the assays to investigate the anti-genotoxic and predict the anti-carcinogenic properties of some medicinal plants extracts with anti-malaria and anti-cancer properties. The ameliorative effect of *Telfairia occidentalis* on radiation induced genetic and systemic damage in rat and the cytogenetoxic effects of its mixture with *Vernonia amygdalina* is being investigated. We are now working on the use of some of the bioassays to study the genotoxic and mutagenic effects of some anti retroviral drugs and metal oxide nanoparticles *in vivo*. The biochemical mechanism of induction of mutagenic and genotoxic effect is also being evaluated.

Recently, I supervised 2 Ph.D theses wherein advanced genetic and molecular techniques were utilized to investigate the mechanisms and also identify the factors of resistance in *Anopheles gambiae sensus lato* and *Anopheles funestus* in Southwestern Nigeria and Southern Benin Republic. Furthermore, I have conducted studies on the genetics of taste using Phenylthiocarbamide taste strips; sex ratio, incidence of twinning and the gene frequencies of the ABO and Rh blood groups and hemoglobin variants in some Nigeria sub-population. As an Alexander Von Humboldt George Forster Experienced Researcher Fellow in Germany, I participated in an investigation into the molecular mechanisms underlying apoptosis induction and cell cycle modulation by isothiocyanates (ITCs) obtained from cruciferous vegetables in hepatocellular carcinoma (HCC) and non-transformed liver cells. We are currently studying the presence, types and concentrations of Pharmaceuticals and Personal Care Products (PPCPs) in South-west Nigeria.

As a University Professor, I teach undergraduate and postgraduate courses in my Department and supervise research projects at the B.Sc., M.Sc. and Ph.D levels in different areas of Genetics; cell, molecular and environmental Biology. I am actively involved in mentoring of junior colleagues in my Faculty. I have served in several administrative capacities in my home Institute; and presently, as the Director of Special Duties in the Vice Chancellor's office, University of Ibadan, Ibadan Nigeria.

Professional Appointments:

1. Graduate Assistant, 01 February, 1994 - Ladoke Akintola University of Technology, Ogbomoso. Nigeria.
2. Assistant Lecturer, 08 June, 1994 - Ladoke Akintola University of Technology, Ogbomoso. Nigeria.
3. Lecturer II, 01 October, 1997 - Ladoke Akintola University of Technology, Ogbomoso. Nigeria.
4. Lecturer II, 02 January, 2001 - University of Ibadan, Ibadan Nigeria.
5. Lecturer I, 01 October, 2004 - University of Ibadan, Ibadan Nigeria.
6. Senior Lecturer, 01 October, 2007 - University of Ibadan, Ibadan Nigeria.
7. Associate Professor (Reader), 01 October, 2010 - University of Ibadan, Ibadan Nigeria.
8. Associate Professor (Reader) on Sabbatical leave - Olabisi Onabanjo University, Ago-Iwoye, Ogun State, Nigeria from 05 May, 2015 – 22 August, 2016.
9. Professor, 01 October, 2013 to date - University of Ibadan, Ibadan Nigeria.

Honors:

1. Member, Genetic Society of Nigeria (GSN).
2. Member, Nigerian Society for Experimental Biology (NISEB).
3. National Executive Member, Zoological Society of Nigeria (ZSN).

4. Editorial Advisory Board member, NISEB Journal, Nigeria.
5. Journal Secretary and Editorial Board Member, The Zoologist, Journal of Zoological Society of Nigeria (ZSN).
6. Honorary Member, American Biographical Institute's Research Board of Advisors.
7. Member, EUROTOX- European Network for Alternative Testing Strategies in ecotoxicology.
8. Fellow, Institute of Policy Management Development (FIPMD).
9. Editorial Board member, World Journal for Biological Research, Cote d'Ivoire.
10. Member, Society of Environmental Toxicology & Chemistry (SETAC AFRICA).
11. CSIR-TWAS Postdoctoral Fellowship at India Institute of Toxicology Research (IITR), Lucknow, India (August 2005 – July 2006).
12. Award of a prize for best presentation in the oral session of International symposium on Environmental Mutagenesis and Public Health and XXXI annual conference of Environmental Mutagen Society of India, held at National Institute of Nutrition, Hyderabad. India, February 23 – 25, 2006.
13. 2005 - 2007 University of Ibadan/John D. and Catherine T. MacArthur Foundation Grant in the Medium term duration category.
14. 2006 University of Ibadan Senate Research Grant (SRG/FSC/2006/17A) for Individual Category for a study on genotoxicity of analgesics commonly dispensed in Nigeria.
15. Co-Principal Investigator in a 2006 University of Ibadan Senate Research Grant (SRG/FSC/2006/1B) for research group category (PI is Prof. O.A. Ugwumba); for a study on Genetic and morphological characterization of culturable fish species in Nigeria.
16. Alexander Von Humboldt (AVH) George Forster Experience Researcher Fellowship in the Department of Environmental Health Sciences, Institute for Environmental Medicine and Hospital Hygiene, Freiburg im Breisgau, Germany (2010 – 2011).
17. Managing Editor, Journal of Science Research, Faculty of Science, University of Ibadan, Nigeria (ZSN).

18. Alexander Von Humboldt (AVH) Return Fellowship for Research Fellows of the AVH Foundation on a project entitled "Anti-genotoxic and anti-carcinogenic effects of some Nigerian medicinal plants" for a period of 1 year (October 2011-September 2012).
19. National President, Genetic Toxicology Association of Nigeria (GTAN)

Publications from the Last 3 Years:

1. Alabi, O. A., Osifo-Whiskey E., Yadi P., Lawal M. and A. A. Bakare. 2014. Tobacco industry wastewater–induced genotoxicity in mice using the bone marrow micronucleus and sperm morphology assays. Cytologia 79(2): 215–225.
2. Alabi, O. A. and A. A. Bakare. 2014. Cytogenotoxic effects and reproductive abnormalities induced by e-waste contaminated underground water in mice. Cytologia 79(3): 1–10.
3. Yekeen, M. O., Oyeyemi I. T. and A. A. Bakare. 2014. Evaluation of the genotoxic, antigenotoxic and histopathological effects of aqueous extracts of Nymphaea lotus L. on sperm morphology and testicular tissue in mice. In Proceedings of the 38th annual conference of Genetics Society of Nigeria held in Benin City, Edo State from 19 – 23 October, 2014. Pp. 305 – 314.
4. Olaniyan, S. A., Amodu O. K., Yindom L., Conway D. J., Aka P., Bakare A. A. and O. O. Omotade. 2014. Killer-cell immunoglobulin-like receptors and falciparum malaria in southwest Nigeria. Human Immunology 75: 816–821.
5. Alimba C. G., Onajole A. I., A. A. Bakare. 2015. Brain dysfunctions in Wistar rats exposed to municipal landfill leachates. Beni - Suef University Journal of Basic and Applied Sciences. 4:284–290.
6. Atoyebi, S. M., I. T. Oyeyemi, B. A. Dauda and A. A. Bakare. 2015. Genotoxicity and anti-genotoxicity of aqueous extracts of herbal recipes containing Luffa cylindrica (L), Nymphaea lotus (L) and Spondias mombin (L) using the Allium cepa (L) assay. African Journal of Pharmacy and Pharmacology 9(15): 492-499.

7. Bakare A. A., Oguntolu G. O., Adedokun L. A., Amao A. A., Oyeyemi I. T., Alimba C..G., Alabi O. A. 2015. In vivo evaluation of genetic and systemic toxicity of aqueous extracts of Phyllanthus amarus in mice and rats. International Journal of Toxicological and Pharmacological Research, 7 (4), August 2015 – September 2015.
8. Alimba C. G., Ajayi E. O., Hassan T., Sowunmi A. A. and Bakare A. A.. 2015. Cytogenotoxicity of abattoir effluent in Clarias gariepinus (Burchell, 1822) using micronucleus Test. Chinese Journal of Biology, Volume 2015, Article ID 624524, 6 pages. (http://dx.doi.org/10.1155/2015/624524).
9. Oyeyemi, I. T., O. M. Yekeen, P. O. Odusina, T. M. Ologun, O. M. Ogbaide, O. I. Olaleye, A. A. Bakare. 2015. Genotoxicity and antigenotoxicity study of aqueous and hydro-methanol extracts of Spondias mombin L., Nymphea lotus L. and Luffa cylindrical L. using animal bioassays. Interdisciplinary Toxicology, Vol. 8(4): 184–192.
10. Alabi, O. A. and A. A. Bakare. (2015). Perceived public health effects of occupational and residential exposure to electronic wastes in Lagos, Nigeria. The Zoologist, 13: 62-71.
11. Olarinmoye, O. M., A. A. Bakare and O. A. Ugwumba. 2015. Pharmaceuticals as environmental contaminants of concern in Nigeria. In Proceedings of Society of Environmental Toxicology and Chemistry (SETAC AFRICA) 1st Scientific Conference of Central and West Africa Region held at University of Nigeria, Nsukka, Enugu State Nigeria., 28 – 31 January, 2015; Eds. Otitoju, O. and Onwurah I. N. E., Pp. 1-8.
12. Fadoju O. M., O. I. Ogunsuyi, C. G. Alimba and A. A. Bakare. 2015. Genotoxicity and mutagenicity of metal oxides nanoparticles. In Proceedings of Society of Environmental Toxicology and Chemistry (SETAC AFRICA) 1st Scientific Conference of Central and West Africa Region held at University of Nigeria, Nsukka, Enugu State Nigeria., 28 – 31 January, 2015; Eds. Otitoju, O. and Onwurah I. N. E., page 9-14.

13. Alimba C. G. and A. A. Bakare. 2016. In vivo micronucleus test in the assessment of cytogenotoxicity of landfill leachates in three animal models from various ecological habitats. Ecotoxicology, 25(2): 310-319.
14. Onifade, R. S., Alimba, C. G., Adenipekun, C. O. and Bakare, A. A. 2016. White Rot Fungus (Pleurotus pulmonarius) Cultivated on lead contaminated rice straw induced haematotoxicity and lead accumulation in liver and kidney of Wistar rats. J. Drug Metab. Toxicol., 7: 210. (doi:10.4172/2157-7609.1000210).
15. Olaniyan, S. A., Amodu O. K., Bakare A. A., Troye-Blomberg M., Omotade O. O. and Rockett K. A. 2016. Tumour necrosis factor alpha promoter polymorphism, TNF-238 is associated with severe clinical outcome of falciparum malaria in Ibadan southwest Nigeria. Acta Tropica 161: 62–67.
16. Olarinmoye, O., Bakare, A., Ugwumba, O. and Hein, A. 2016. Quantification of pharmaceutical residues in wastewater impacted surface waters and sewage sludge from Lagos, Nigeria. J. Environ. Chem. Ecotoxicol., 8(3): 14-24.
17. Yekeen, T. A., O. O. Fawole, A. A. Bakare and Emikpe B. O. 2016. Alteration in haematological, biochemical and reproductive indices of Rattus norvegicus treated with lambdacyhalothrin. Zoology and Ecology, 26 (1): 47-56.
18. Bakare, A.A., Udoakang, A.J., Anifowoshe, A.T., Fadoju, O.M., Ogunsuyi, O.I., Alabi, O.A., Alimba, C.G. and Oyeyemi, I.T. 2016. Genotoxicity of Titanium dioxide nanoparticles using the mouse bone marrow micronucleus and sperm morphology assays. Pollut. Eff. Cont. 4: 156. doi:10.4172/2375-4397.1000156.
19. Alimba, C. G., D. Gandhi, S. Sivanesan, M. D. Bhanarkar, P. K. Naoghare, A. A. Bakare, K. Krishnamurthi. 2016. Chemical characterization of simulated landfill soil leachates from Nigeria and India and their cytotoxicity and DNA damage inductions on three human cell lines. Chemosphere 164:469-479.

20. Djouaka, R.J., Atoyebi, S. M., Tchigossou, G. M., Riveron, J. M., Irving, H., Akoton, R., Kusimo, M. O., Bakare, A. A. and Wondji, C. S. 2016. Evidence of a multiple insecticide resistance in the malaria vector Anopheles funestus in South West Nigeria Malaria Journal, 2016 15:565 (DOI 10.1186/s12936-016-1615-9).
21. Oyeyemi, I. T., O. O. Akanni, O. A. Adaramoye and A. A. Bakare. 2017. Methanol extract of Nymphaea lotus ameliorates carbon tetrachloride-induced chronic liver injury in rats via inhibition of oxidative stress. J. Basic Clin. Physiol. Pharmacol., 28(1): 43–50.

INDEX

C

D

N

O

P

R

S